Rocks, Minerals, and Crystals

D. C. Almond PhD
D. G. A. Whitten

illustrated by Max Ansell
and John Smith

Hamlyn
London · New York · Sydney · Toronto

FOREWORD

Interest in geology is increasing at a considerable rate partly due to the growing realization that so much depends upon our knowledge of the Earth – energy supplies, raw materials, water supplies, soil conservation, and pollution problems come to mind immediately. A major source of information for the geologist is in the material making up the crust of the Earth – the rocks we see around us. It is the aim of this book to provide a background of basic principles for the study of rocks and minerals: it is *not* an identification manual, rather it is an introduction to the methods and ideas of the geologist.

Geology is one of the few areas of science where the amateur can still make a significant contribution. Perhaps the lure of geology is the lure of the unknown, the realization that there is something new to find out just over the next hill – and it is there for anyone to find. Geology has a fascination all its own, and many people first come to appreciate this through collecting rocks and minerals. If this book helps its readers to a fuller understanding of their collections, and produces a desire to learn more about them, it will have succeeded. Good hunting!

DCA
DGAW

Published by the Hamlyn Publishing Group Limited
London · New York · Sydney · Toronto
Astronaut House, Feltham, Middlesex, England

ISBN 0 600 31873 7

Phototypeset by Filmtype Services Limited, Scarborough, England
Colour separations by Metric Reproductions Limited, Chelmsford, England
Printed in Spain by Mateu Cromo, Madrid

CONTENTS

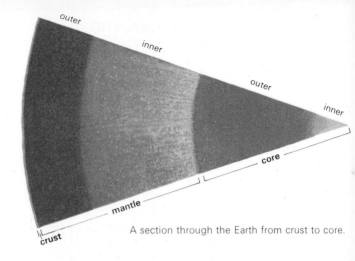

A section through the Earth from crust to core.

THE STRUCTURE OF THE EARTH

The Earth that we see and make contact with is largely solid: the huge volume of the oceans (about 1300 million cubic kilometres) makes up only a minute fraction of the Earth's total volume (more than a million million cubic kilometres). This solid Earth is made up of **rocks**, which in turn are made of simpler units called **minerals**. With the exception of meteorites and the Moon rock samples the only material available for study is that which can be obtained at or reached from the Earth's surface. The deepest borehole yet drilled (in Texas) has a depth of 19 kilometres. The deepest mine is in South Africa and is 3·4 kilometres.

The mean diameter of the Earth is 12 742 kilometres. Thus, man has barely penetrated more than a small fraction of 1 per cent of the Earth's diameter. Yet, the geologist *must* attempt to understand the interior of the Earth if he or she is to make sense of the outer layers. This understanding is vital because it is upon these outer layers that we depend for all our fuels, water, and raw materials. We can directly observe about 16 kilometres of the uppermost layer, the **crust**, because of the way rocks have been uplifted from the depths, tilted, and dissected. By using sensitive instruments and making reasonable assumptions, geologists have constructed a 'model' of the Earth's interior. The study of thousands of records of earthquake shock waves which have passed through the Earth and are modified (slowed or accelerated) by the various materials they traverse and from other evidence as well, yields the 'model' shown on page 4. The central **core** of nickel-iron probably produces the Earth's magnetic field, while the **mantle** which forms the bulk of the Earth may be the indirect source of activity and material in the uppermost portion, the crust.

The crust is about 30 kilometres thick, although it varies considerably, being thinnest under the oceans and thickest under mountain ranges. Recent work has yielded the model of the crust shown on page 5. The crust is mobile at certain times and in certain places, that is, it deforms and moves, and allows material from the deeper parts to move towards the surface, sometimes even to permit extrusion at the surface.

Despite the difficulties, a systematic procedure for describing, identifying, and classifying minerals and rocks has evolved. More advanced techniques have improved this. It is still possible, however, to gain a considerable insight into the basic characteristics of minerals and rocks without access to equipment more complex than a hammer, a cold chisel, a compass, notebook, maps, and guides. A few hours spent prospecting around a quarry, cliffs, or an old mine tip will start you off. Before long you will have the nucleus of a collection (and probably a storage problem!) you will realize the need to read more, and to see more.

A visit to the Mineral Gallery of the Natural History Museum will be a revelation of the wonderful specimens which have been found in the past – and are still being found by people who are prepared to search thoroughly and carefully. Many other museums have interesting displays of rocks and minerals, often from their local area, and may publish guides. If you need help, do not be afraid to ask: geologists are always ready to assist a genuine enthusiast.

A section through the Earth's crust.

mountain belt

ocean

continental crust

oceanic crust

mantle

COMPOSITION OF THE EARTH

By means of detailed chemical analysis of many materials from the accessible parts of the Earth's crust, a picture emerges of the broad average composition of the crust. Man relies on this part of the Earth for all his raw materials of 'mineral' origin, so that it is obviously most important that a clear idea of the distribution of the individual chemical elements is obtained. The result of these analyses is rather surprising. Oxygen makes up nearly half the crust (47 per cent), and the next commonest element is silicon (27 per cent) so that the most likely way for these two elements to occur is in combination in the form of a **silicate**. The six next most abundant elements (aluminium, iron, calcium, sodium, potassium, and magnesium) total 24 per cent, so that eight elements constitute 98 per cent of the Earth's crust. There are about ninety elements known in the crust so that eighty or so elements make up the remaining 2 per cent. See page 7.

Many elements which we regard as 'common' must actually occur in the Earth's crust in very small quantities. For example, four everyday metals, copper, lead, tin, and zinc occur to the extent of 45, 16, 3, and 65 parts per million respectively. Other elements are even scarcer: mercury occurs at only 0·5 parts per million while silver is only 0·1 parts per million.

We must now consider the Earth as a whole and as part of the Universe. Astronomical evidence, especially the study of meteorites, suggests strongly that, considering the whole Earth, the four most important elements, in order of abundance, are iron (40 per cent), oxygen (28 per cent), silicon (14·5 per cent), and magnesium (8 per cent); nickel and sulphur (2·5 per cent each) are next in order (their crustal abundances are 80 and 520 parts per million respectively). From what has been said about core and mantle, these proportions are to be expected, and they fit the model of the Earth we have built up. In the whole Universe, hydrogen and helium are far more abundant than any other elements – hydrogen being the most abundant. The stars and planets represent minute 'specks' where other elements are concentrated.

We must now turn back to the Earth's crust. If elements such as copper and lead are so scarce, how are we able to extract them for use? The answer lies in the first statement in this section – the abundances of the elements are averages, that is, as if the element were spread evenly through the whole crust. This can easily be seen to be untrue because the materials comprising the crust are obviously not uniform in character, and these materials do display considerable variation in appearance and composition.

Processes operating in the crust have caused some elements to become locally concentrated to such an extent that we can use them. For instance, a workable, lead-bearing vein may contain 2 per cent of the metal, representing a concentration factor of 1250, that is, there is 1250 times as much lead in the vein than there is on average in the crust. A 1 per cent tin vein involves a concentration of 3330 times. Not all elements achieve high local concentrations and we are just beginning to realize that some of these are useful and significant. For example, germanium, the vital element in the transistor, is never found in high concentrations and is extracted as a by-product when found in association with other elements.

Proportions of the elements in the Earth's crust.

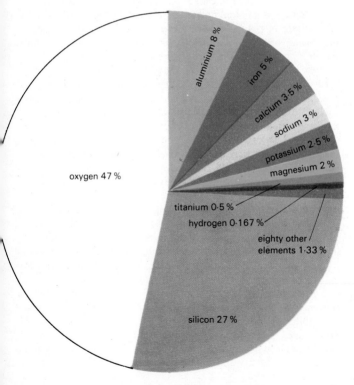

MINERALS AND ROCKS

You have seen that the elements are not distributed uniformly in the crust, and we can demonstrate this by examining in detail the rock, granite, which is common in many areas. In Britain granite is found in south-west England and the northern parts of Scotland. Polished slabs are often used as facing stone in large buildings. If we examine this material closely using a hand-lens, we will find several different substances which make up granite. Firstly, a greyish, rather glassy-looking substance, which is very hard; secondly, a white, grey or pinkish material, with an appearance not unlike china, fairly hard, often in rectangular tablets; thirdly, a much softer, flaky substance, either a pale, silvery colour or dark brown to black.

Chemical analysis shows that these three substances are compounds with fixed compositions and are not mixtures of simpler substances. In fact, the first material consists of silicon and oxygen – silicon oxide – which is called quartz, the second is a silicate of aluminium, sodium, and potassium, called feldspar, while the third is a silicate of potassium and either aluminium (the white mica, muscovite) or magnesium, iron, and aluminium (the dark mica, biotite). These compounds are termed minerals and the association of these minerals in granite is called a rock. A mineral can be defined formally as: *a homogeneous, solid substance, having a fixed chemical composition, formed by the inorganic processes of nature.* Thus, we exclude all man-made materials, water, the atmosphere, oil and gas, and composite materials like granite. A rock can be defined formally as: *an aggregate of separate mineral units.* Notice that this means that a rock may be made up of a single mineral or a number of different minerals and that rocks do *not* need to be hard and compact – the loose sand on a beach is as much a rock to the geologist as is granite.

Some of the material which makes up certain rocks may be derived from plants or animals and, although strictly this kind of material is not mineral, if it is identical to a true mineral, it is generally admitted as such. For example, the plates forming the skeleton of a sea-urchin are made of calcium carbonate, identical with the mineral calcite.

You can see that a rock, made up of several different minerals, may vary in composition as the relative proportions of the constituent minerals vary. Thus, in a granite, almost all the iron in the rock is contained in biotite, and the amount of biotite present is roughly proportional to the amount of iron present. Clearly, it is important to be able to identify and describe minerals accurately because it is only then that we can distinguish the various kinds of rocks, each of which is defined in terms of the minerals present. (Other properties are also used and are described later.) Beside the **essential minerals** present in a rock, very small amounts of what are called **accessory minerals** may be present. Sometimes, minerals become altered to

A rock is formed of minerals.

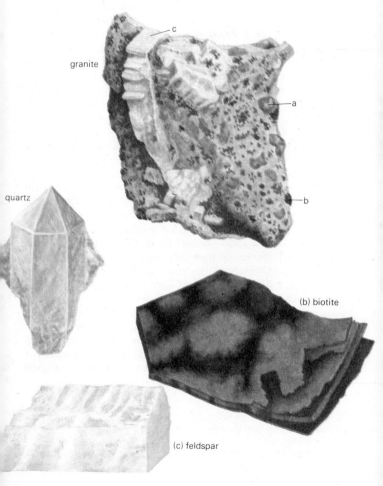

granite

quartz

a

b

c

(b) biotite

(c) feldspar

other minerals, and these are referred to as **alteration products.**

The rock granite which we have been examining is typical of rocks which have solidified from a molten condition. When the lava from a volcano solidifies it yields other varieties of such rocks, which are termed **igneous**. Almost all the constituent minerals of igneous rocks are silicates (including quartz).

When rocks are exposed at the Earth's surface, they are attacked by the rain and atmospheric conditions, broken up, and carried away by rivers and streams, the wind, the sea, and even glaciers. This derived material is eventually deposited somewhere (usually the sea) and forms **sedimentary rocks**. Silicates (including quartz) and carbonates make up a high proportion of the minerals of these rocks.

Under certain circumstances rocks may be subjected to intense heat and/or pressure, resulting in the reorganization of the rock, including the development of new minerals. These rocks are termed **metamorphic**. They are essentially made of silicate minerals (including quartz) although carbonates do occur.

Rocks consist mainly of silicate minerals, but non-silicate minerals containing the 'rare' elements occur locally and sometimes in quantities sufficient for them to be exploited. A patient search of the dumps of waste material from old mines will frequently yield interesting specimens, while quarries and sea-cliffs may also be productive. Britain abounds in old mines and when visiting an area to search for minerals, you should make a preliminary survey, using the appropriate Regional Geology Guide. You may often obtain more detailed information from local museums and libraries, while the larger scale Ordnance Survey maps usually indicate the position of old mines and quarries. Many of them will be on private land, so that permission for a visit is nearly always necessary. Old mining areas are often dangerous: concealed shafts may exist, and under no circumstances should you enter old shafts or tunnels. Disused quarries can also be dangerous, while working quarries may or may not be willing to admit visitors. Finding minerals is often a matter of luck, although experience will guide you to the more likely parts of an exposure.

Some typical rocks: (a) fine-grained, volcanic, igneous; (b) coarse-grained, deep-seated, igneous; (c) fragmental sediment – pebble bed; (d) organically formed sediment – fossiliferous limestone; (e) strongly metamorphosed, showing banding; (f) recrystallized, metamorphosed limestone.

THE GEOLOGICAL CYCLE

The study of rocks in detail involves investigation of their relationships, and this in turn has revealed a good deal about the way the chemical elements behave in the Earth's crust and the mineralogical consequences of such behaviour. If we follow some rock material from its origin deep in the crust or upper mantle, we can establish some useful principles which help to explain rock and mineral relationships. The illustration will help to make things clearer.

Commencing with a rock-melt (**magma**), formed deep down, we may find that this has solidified either after moving upwards to higher levels in the crust (**intrusive**) or after flowing out at the surface (**extrusive**). The resulting igneous rocks will vary in composition (both chemically and mineralogically), partly due to initial differences and partly due to differences of cooling history during the upward passage through the crust towards the surface. If the original melt contained not only the 'common' elements but the normal proportion of 'rare' ones, we may find that during the solidification process these 'rare' elements become concentrated and form separate masses, associated with larger masses of more normal igneous rock. For example, the large granite mass of Dartmoor, in Devon, is a normal igneous rock, while around it we find small masses of minerals containing lead, zinc, copper, tin, and arsenic which have been derived from the original granite magma.

The igneous rocks will ultimately become exposed to the atmos-

The geological cycle.

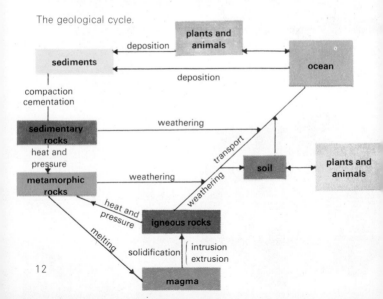

phere, as a result of weathering processes, and will themselves be broken down by atmospheric agencies. Some of this weathered material becomes soil, where its constituents are available to feed organisms which, when they die, return their substance to the soil. Much of the weathered material (and some soil) is transported by rivers to the sea, however: some material is in solution, some in suspension. The suspended material deposits on the sea floor to form sediments, while the material in solution may do one of three things:

1 remain in the sea as, for example, sodium and chlorine tend to do; 2 be extracted by plants or animals; for example, calcium is extracted to form animal skeletons (shells, coral, and so on) in the form of calcium carbonate, which is usually incorporated in the sediments, and other substances extracted may be returned to solution; 3 be precipitated chemically; for example, iron and manganese oxides and some calcium carbonate are probably deposited in this way. Sediments are generally converted to a sedimentary rock by compaction and/or cementation.

At this point there are two possible courses that may be followed. The sedimentary rock may be weathered and transported back to the sea, and this cycle may be repeated many times. Alternatively, a sedimentary rock may be altered by metamorphism into a metamorphic rock. This may then be either weathered and transported, re-entering the sedimentary part of the cycle or it may be heated and compressed to such an extent that it melts or is assimilated by magma to form new igneous rock material. At each stage in the cycle, some elements are concentrated, others dispersed, giving rise to the mineralogical variations found in rocks.

CRYSTALS

An essential property of most solid substances is **crystallinity**. The word, **crystal**, was first applied to the mineral quartz: the clear, glassy prisms of the material led the ancients to believe that they were 'permanent' ice crystals – *krystallos* – and the name, crystal, spread to include any regularly shaped unit of a mineral. Later, it was discovered that, in all crystals of the same substance, the angles between corresponding pairs of faces were constant and characteristic of the substance. We now know that this is a result of the atoms in the crystal being arranged in a fixed, regular pattern, known as an **atomic lattice**.

The atomic lattice is present in a crystal fragment, just as much as it is in a well-formed crystal. The term, **crystalline**, implies a regular atomic structure, while the term, **crystallized**, implies a well-formed crystal. Sometimes fine-grained (ultramicroscopic) aggregates occur, which have some crystalline properties; these are termed **cryptocrystalline**, the individual units being **crystallites**. Non-crystalline material (which is rare among minerals) is termed **amorphous**: many so-called amorphous minerals show slight traces of crystallinity.

The regularity of the atomic lattice also leads to the property of crystal symmetry – the main property upon which crystals are classified. We use the idea of symmetry frequently in everyday life, often unconsciously: for example, left hand – right hand is a symmetry-based idea; four players seated round a square card-table is another. Three sorts of symmetry can be defined formally.

1 **Planes of symmetry** divide solids into two 'mirror-image' halves (left hand – right hand).

2 **Axes of symmetry** are lines about which rotation can occur so that the same position of the crystal is repeated several times per rotation. Only 2-, 3-, 4-, and 6-fold axes are possible. The four players at the card-table are arranged around a 4-fold axis.

3 **Centre of symmetry** is a point through which lines connecting

Crystallographic symmetry: (a) plane of symmetry; (b) axis of symmetry (3-fold); (c) centre of symmetry.

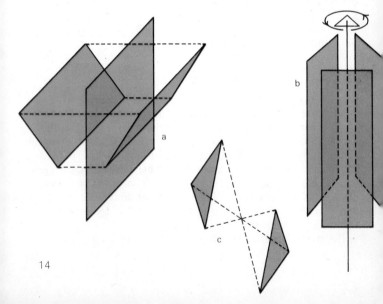

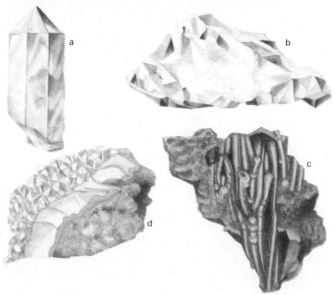

Forms of silica: (a) crystallized, a single crystal; (b) crystalline, a crystal fragment; (c) cryptocrystalline chalcedony; (d) non-crystalline (amorphous) opal.

equivalent points in the crystal pass. Its main effect is to produce pairs of parallel faces.

Using the basic symmetry units, mathematicians have shown that thirty-two classes of symmetry are possible: a number are unknown or very rare in nature. The thirty-two classes can be grouped into seven systems: cubic, hexagonal, trigonal, tetragonal, orthorhombic, monoclinic, and triclinic.

The faces which make up a crystal occur in sets, known as **forms**; the number of faces comprising a form depends on the symmetry. For descriptive purposes, faces are related to sets of axes, called **crystallographic axes**, not to be confused with symmetry axes, although the two may coincide. Six sets of axes are used for the seven systems, the hexagonal and trigonal systems sharing a set consisting of three axes arranged at 120 degrees to one another in a horizontal plane, with a fourth perpendicular to them. The other systems use three axes: in the cubic, tetragonal, and orthorhombic systems they are mutually perpendicular, while in the monoclinic system one axis is not at right angles to the other two, which are at right angles. In the triclinic system, no two axes are at right angles. For reference purposes, the vertical crystallographic axis is labelled c, the left-to-right one b and the back-to-front one a. Faces are named according to which axes they cut. For example, a face cutting all three axes is termed a **pyramid**; cutting two horizontal axes and parallel to the vertical axis, a **prism**; cutting one horizontal axis and parallel to one vertical and one horizontal axis, a **pinacoid**. The relative development of different faces controls the **habit** of a crystal. For example, a flat, 'slabby' crystal is termed 'tabular', while one that is elongated is termed 'prismatic'.

The forms developed in the cubic system are special and sometimes complex because of the high symmetry: they are given special names, such as cube, octahedron, rhombdodecahedron, pyritohedron, which are not used in other systems.

ATOMIC STRUCTURE

Atoms are linked together in regular ways which may be simple or complex. Some atoms acquire a positive electrical charge, others a negative one. Positively charged atoms (ions) are termed cations, negatively charged ones, anions. In a crystal, the positive and negative charges bind the atoms together, and cancel out, leaving an uncharged crystal. A simple crystal, such as rock salt, has atoms organized in a cubic array, that is, with lines of atoms, alternately sodium (positive) and chlorine (negative), arranged in three mutually perpendicular directions, parallel to the edges of a cube, the whole making up a cubic lattice. The repeat unit of an atomic structure is called a **unit cell**.

Silicon (positive) and oxygen (negative) build a whole series of complex lattices which are very important in our understanding of the rock-forming silicate minerals. Basically, one silicon atom links to four oxygen atoms in a very stable tetrahedral arrangement (SiO_4). In some minerals, SiO_4 units are packed closely together, with magnesium and iron cations, yielding a structure, which is compact and dense.

SiO_4 units can also be linked together in chains, each SiO_4 unit sharing two of its oxygen atoms, one with each neighbouring unit. This chain has effectively the composition, SiO_3, and occurs in the group of minerals known as the **pyroxenes**. The chains are packed together with appropriate cations (calcium, magnesium, iron,

The crystal systems.

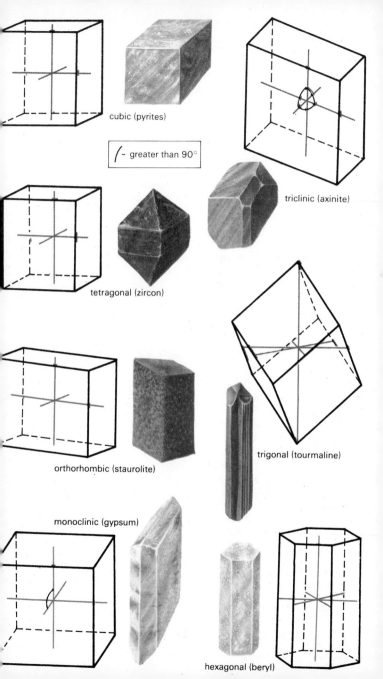

cubic (pyrites)

/ - greater than 90°

triclinic (axinite)

tetragonal (zircon)

orthorhombic (staurolite)

trigonal (tourmaline)

monoclinic (gypsum)

hexagonal (beryl)

aluminium) parallel to the c-axis of the crystal. The binding between the chains is weaker in some directions than others, so that two planes of weakness arise parallel to the prism faces of the crystal, along which the mineral splits or **cleaves**.

Sometimes, chains of SiO_4 units link up to form rings – three-, four-, and six-membered rings are known.

The SiO_3 chains can also link up along their length to give 'ribbons' (as in the **amphiboles**), which also possess a prismatic cleavage, and these in turn may cross-link into sheets or layer lattices (as in the **micas**) which have a cleavage, parallel to the layers, yielding thin leaves of the mineral. Both the 'ribbons' and the sheets also include oxygen-hydrogen units, called hydroxyl (OH), which are part of the water molecule.

The most complex silicon-oxygen structures are the framework or three-dimensional lattices, found in **quartz** and the **feldspars**. Several other silicon-oxygen lattices exist, some in rare and unusual minerals.

It is noticeable that the main cations present in silicates are sodium, calcium, magnesium, iron, aluminium, and potassium while other atoms are present only in very small amounts. This is because the atoms listed above are the commonest atoms, after silicon and oxygen (p 7) and also they are the right 'size' to fit into the 'holes' in the lattices. The latter is a vital factor because many elements have atoms which are either too big or too small for silicate lattices and even the common elements are not found in all lattices. Potassium, for example, is mainly found in the layer lattices and the framework lattices.

An understanding of these principles has greatly advanced our knowledge of minerals, particularly in establishing a classification and in explaining physical properties. Determination of atomic structure involves powerful X-ray generators and is beyond the capabilities of most people. Every time you examine a crystal and observe its cleavage and other properties, and look at it in thin section, you are experiencing the consequences of its special atomic structure.

HOW MINERALS ARE STUDIED

A geologist collecting minerals for study needs fresh, unweathered material and the best specimens to be found. He or she begins by

Atomic lattices: (a) simple cubic; (b) pyroxene chain – top view; (c) pyroxene chain – end view; (d) pyroxene chains packed in crystal.

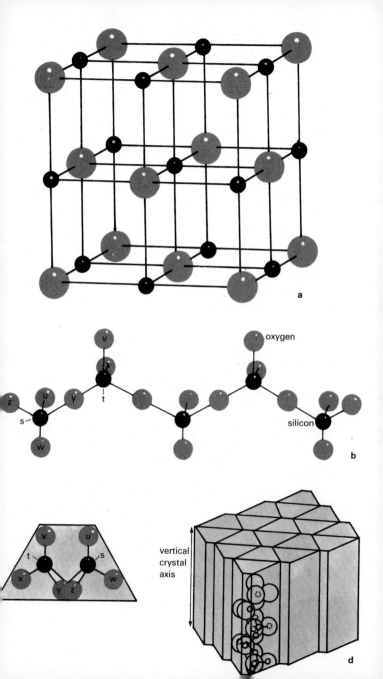

oxygen

silicon

vertical
crystal
axis

carefully examining the locality, sketching or photographing it if necessary. Thus, specimens can be accurately positioned and their relationships studied. Even if specimens are being picked up from a mine tip, this procedure is followed. Hammer and chisel are used cautiously – it is easy to break crystals merely by the shock of hammer blows. All the fragments should be collected, even broken crystals being useful, especially if the mineral being extracted is uncommon. Specimens are carefully labelled, wrapped, and packed for transport to the laboratory: paper, cotton wool, polythene bags, and boxes are essential. Detailed examination starts with a thorough inspection through a hand-lens. For very small crystals, a binocular microscope is used.

The geologist can now use four main techniques of study. Two of these, chemical analysis and X-ray studies, are not easily available to the amateur but are essential in full studies of minerals. Some simple chemical tests are feasible for the amateur. Ten per cent hydrochloric acid is an infallible test for carbonates, for example, (dolomite needs *warm* acid) producing a characteristic effervescence. Other simple chemical tests can be found in *Rutley's Elements of Mineralogy*. The other two techniques are the main ones available to the amateur.

A technique that will be discussed later involves the use of the polarizing microscope. Minerals and rocks to be studied in this way have to be made into thin, transparent sections with a thickness of 0·03 millimetres which involves extremely careful working. The starting point for making thin sections can either be a slice, 2 or 3 millimetres thick, cut by means of a diamond saw or a flake produced when hammering. This is then ground down to a smooth finish, mounted on a glass slip by means of Canada balsam (or a synthetic resin) and finished to the required thickness. Grinding is done by means of successively finer grades of abrasive. Great care is necessary in the final stages to avoid completely grinding away the thin slice. The remaining technique is so important and so widely used that the next section is devoted to it.

THE PHYSICAL PROPERTIES OF MINERALS

A geologist nearly always starts to identify a mineral by first using a hand-lens, then trying to scratch it with a knife, and finally 'weighing' it in the hand. Experienced workers can use the various physical properties of the specimen to identify it.

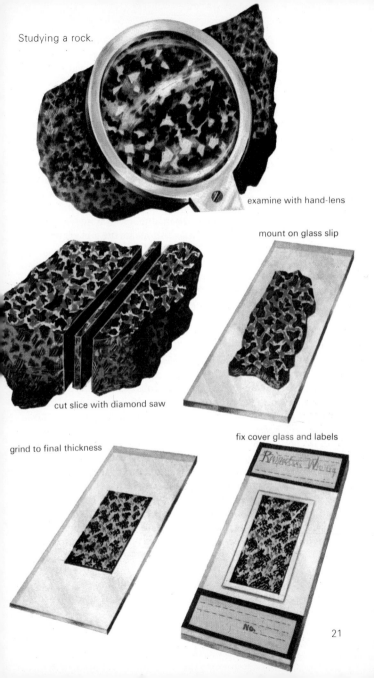

Studying a rock.

examine with hand-lens

mount on glass slip

cut slice with diamond saw

grind to final thickness

fix cover glass and labels

21

Visual properties

Colour is an obvious property but unfortunately it is one of the least reliable tests available. Many minerals such as quartz and fluorite occur in a variety of colours while very few minerals have a constant colour – malachite and sulphur are examples. The colour of the powdered mineral, the **streak**, is more useful. This is obtained by scratching the mineral on a piece of unglazed porcelain, a streak plate, or by using a file and catching the powder on white paper. The streak may be the same colour as that of the mineral; it may be coloured but different from the colour of the mineral; the mineral may be coloured and the streak white; and finally, the mineral may be too hard to produce a streak. Care must be taken to distinguish between a white streak and no streak. This test is useful in distinguishing some of the black minerals; for example, haematite gives a reddish-brown streak, limonite a yellowish-brown one, magnetite a black one, and cassiterite no streak.

The way in which a mineral reflects light, its **lustre**, is often characteristic. Many sulphides, for example, have a metallic appearance whereas many silicates have a glassy (vitreous) lustre. Other lustres are described as resinous, pearly, earthy, and so on.

Hardness, fracture, and cleavage

The most important simple identification test is hardness, determined by scratching the unknown mineral with one of known hardness, belonging to the standard Mohs' scale of hardness. This consists of ten minerals, numbered 1 to 10 in increasing order of hardness, as follows: talc 1; gypsum 2; calcite 3; fluorite 4; apatite 5; orthoclase feldspar 6; quartz 7; topaz 8; corundum 9; diamond 10.

In the absence of standard samples, alternatives may be used; for example, a fingernail is about $2\frac{1}{2}$, a 'copper' coin about 5, window glass $5\frac{1}{2}$, and a penknife $6\frac{1}{2}$. These enable pairs of similar minerals to be distinguished quickly. For example, pyrite has hardness 6, while the similar chalcopyrite has hardness 4: a penknife barely scratches the former but easily scratches the latter. When testing hardness, make sure the unknown specimen is really scratched – powder may be produced from the scratching specimen. Granular specimens may

Properties of minerals: (a, b, c) purple, colourless, and green fluorite; (d) streak plate; (e) metallic lustre; (f) resinous; (g) vitreous or glassy; (h) single cleavage; (j) three cleavages at right angles; (k) conchoidal fracture.

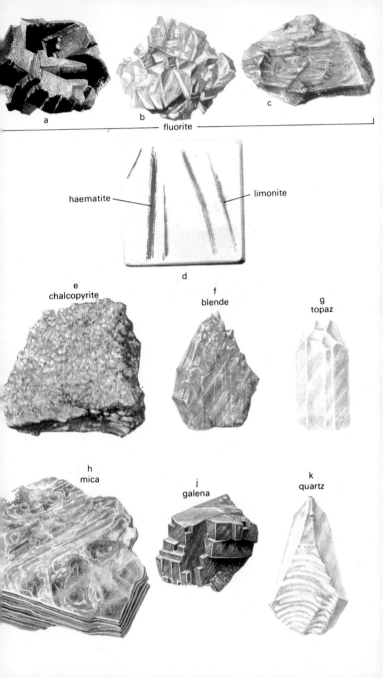

a

b

c

fluorite

haematite — limonite

d

e
chalcopyrite

f
blende

g
topaz

h
mica

j
galena

k
quartz

be disintegrated by the test specimen without the grains themselves being scratched.

Any piece of calcite (preferably not a good crystal) tapped firmly, will break into small rhombohedra and if these are powdered the grains will still be the same shape. It is, in fact, quite difficult to break calcite except into rhombohedra. These three directions are known

Mineral aggregates: (a) platy, 'cock's comb'; (b) fibrous; (c) radiating; (d) reniform or kidney shaped.

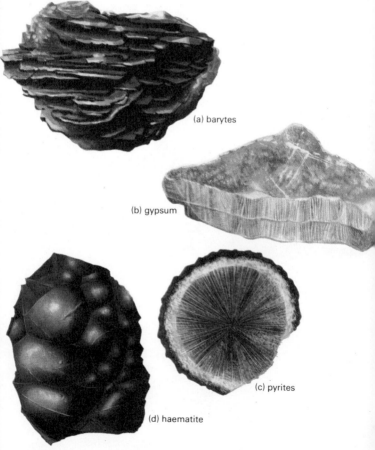

(a) barytes

(b) gypsum

(c) pyrites

(d) haematite

as **cleavage planes** and are due to weaknesses in the atomic structure in these special directions (which are possible crystal faces, although not all crystal faces are cleavage planes). Examples of cleavage can be seen in micas, which split easily into thin sheets, and galena, which breaks into cubes. Quartz, however, behaves differently; when struck with a hammer it fractures like a piece of glass. Several minerals fracture without cleaving, some irregularly, like metal, others, like quartz, showing the curved 'shell'-like form known as **conchoidal fracture**.

State of aggregation

The individual crystals of a mineral may be packed together in such a way that the individual shape of the crystals is lost in a characteristic aggregate. For instance, gypsum crystals are normally flat (p 47). When they grow in a fissure, they commonly develop a fibrous structure. Close inspection will reveal individual 'needles' of gypsum.

Many minerals occur in masses of needle-like crystals, some being parallel aggregates, others radiating from a centre; for example, asbestos, pyrite, and calcite. Masses of platy crystals are also common, some being rather open like the cock's comb variety of barytes, others being much more compact and massive, like the kidney-shaped mass of haematite. Other aggregates occur; for example, granular magnetite is common. A mineral which commonly occurs in a typical aggregate is often easy to identify, but remember that it may not always occur in this special form.

Other properties

Some minerals have highly characteristic properties which can distinguish them from almost all others. For instance, the only common mineral with a salty **taste** is rock salt; the only common, strongly **magnetic** mineral is magnetite (p 43). A useful skill is to be able to estimate specific gravity, that is, the relative weight compared with an equal volume of water. We can divide minerals into very heavy, heavy, moderate, medium, light groups by 'hefting' specimens in the hand, provided the specimens are about the same size. As a guide, galena (7·5) can be regarded as very heavy, pyrites (5) as heavy, fluorite (3) as moderate, feldspar or quartz (2·6) as medium, gypsum (2·3) as light. For accurate determination of specific gravity, refer to the list of books to read at the end.

GEMSTONES

Many minerals have been cut and polished to display their beauty, mainly for personal adornment. The four most important (and valuable) gems are diamond which is crystalline carbon; ruby and sapphire which are red and blue corundum, an aluminium oxide; and emerald, a brilliant green variety of beryl, a beryllium aluminium silicate. All these minerals share some common features: they are all very hard – diamond 10, ruby and sapphire 9, beryl 8; they all occur in brilliant, transparent crystals which sparkle when cut and polished due to the internal reflection and refraction of light; they possess no strong cleavage, which would make them fragile and difficult to cut. Other minerals have these properties; olivine (peridot), tourmaline (many coloured varieties), topaz (orange, pink, blue, but has a cleavage), aquamarine (pale blue beryl), garnets (red) are examples and produce some beautiful stones. They are judged to be 'commoner' than diamonds, rubies, sapphires, and emeralds, however, so that they are not as highly prized. Other 'common' minerals sometimes yield semiprecious stones. Many quartz varieties are cut; for example, amethyst (violet), citrine (yellow), cairngorm (brown), crystal (colourless). Occasionally, unusual minerals have been cut, such as cassiterite, transparent pieces of which are extremely rare, or fluorite, despite its softness and cleavage, and zircon, zirconium silicate, which occasionally yields blue or colourless stones.

All the stones mentioned are transparent, but one or two precious

Gemstones: (a) crystal and brilliant cut; (b) crystal and oval cut; (c) crystal and cabochon cut; (d) crystal and emerald cut.

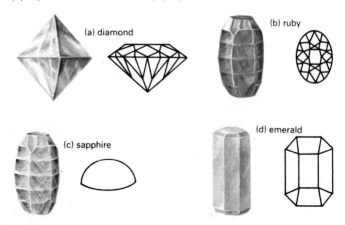

(a) diamond

(b) ruby

(c) sapphire

(d) emerald

and several semiprecious stones are partly or wholly opaque. Some material has a kind of fibrous structure, which produces the interesting effects seen in such materials as cat's-eye, tiger's-eye, and star sapphires.

Banding of different colours makes agates and their varieties interesting, while the deep, solid colour of carnelian, jasper, onyx, and bloodstone is very attractive. Opal is in a class of its own. The brilliant play of colours is purely an optical effect produced by interference in a mass of microscopically thin films present as inclusions in the opal in a similar way to the colours produced by a thin film of oil on the surface of water. Many gemstones are extracted from river gravels, having been weathered out of their parent rocks (igneous or metamorphic) and thence into streams.

CLASSIFICATION OF MINERALS

The classification of minerals is based largely on their chemical composition, and to some extent on their atomic structure. A commonly used classification is shown below, with examples of minerals noted in brackets.

SILICATES

single SiO_4 groups (olivine, garnets)

ring silicates (tourmaline, beryl)

single chains pyroxenes (augite)

double chains or ribbons amphiboles (hornblende)

sheets (micas, chlorite, talc, clay minerals)

three-dimensional structures
feldspars (orthoclase)
feldspathoids (nepheline)
silica group (quartz)

miscellaneous silicates
(aluminium silicates, zeolites, topaz, epidote)

NON-SILICATES

native elements (copper, graphite)

oxides (magnetite, corundum)

sulphides (pyrite, galena)

halides (rock salt, fluorite)

oxygen salts carbonates (calcite)
sulphates (barytes)
phosphates (apatite)
borates (borax)

Strictly, the silica group belongs with the oxides but is more conveniently grouped with the silicates.

THE OCCURRENCE OF MINERALS

The occurrence of minerals and the occurrence of rocks are essentially the same – and because rocks are everywhere, minerals are, too. The sand on the beach is the mineral quartz in fine particles. A big limestone quarry is mainly the mineral calcite. The metamorphic rock that builds a Scottish mountain is made of quartz, mica, and perhaps garnets. The common rocks are made of a limited range of minerals, however, mainly silicates (calcite, calcium carbonate, is the main exception), and many well-known minerals do not occur as essential rock formers.

It is in the group of non-rock-forming minerals that most of the major economic minerals are found, so that we must now consider some of the ways in which these occur. The term, **mineral deposits**, is commonly used to describe such occurrences, regardless of their actual modes of formation.

Mineral deposits associated with igneous rocks

When a large igneous mass cools and solidifies, a small proportion is left which differs essentially from the main mass of rock. For example, a granite consisting of quartz, feldspar, and mica may produce a fraction rich in tin, copper, lead, zinc, and so on because these elements can not find a place in the structure of quartz, feldspar, or mica. Thus, the tin, copper, lead, and zinc atoms will become concentrated in the last residue, together with elements such as sulphur, fluorine, iron, barium, and calcium in a liquid rich in water and carbon dioxide.

This liquid is hot and is probably midway between a melt rich in water and a highly concentrated watery solution: it is called a **hydrothermal fluid** and is responsible for a variety of mineral deposits. The simplest form is the **vein**, a sheet-like mass of minerals cutting through the rocks often occupying joints and fissures. Veins vary from paper-thin veinlets to bodies tens of metres thick and hundreds of metres long. Sometimes veins occur in great numbers, in parallel sets, as radial systems, or as a random, mesh-like pattern.

Two sorts of mineral occur in such veins: **ore minerals** such as sulphides and oxides, and **gangue minerals** such as quartz, calcite, barytes, and fluorite. The word, gangue, implies waste or worthless and was originally used by miners. Nowadays, however, gangue minerals such as fluorite and barytes are far from worthless and may be actively sought and mined.

Some veins consist of large crystals of rock-forming silicates, together with crystals of rarer minerals such as beryl, lithium and uranium minerals, and ore minerals such as cassiterite (tin oxide). These veins are called **pegmatites** (see later) and occasionally contain crystals several metres long.

Sometimes hydrothermal fluids replace existing bodies of rock – usually limestones and form large bodies of minerals. Lead and zinc ores are often found in this form. Often, the mineralizing fluids also metamorphose the rocks, producing an association of metamorphic and ore minerals called **skarns**.

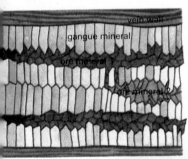

Top Section through a vein.
Centre Veins and replacement bodies associated with an igneous mass.
Bottom Placer deposits.

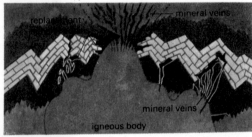

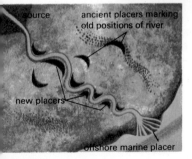

Mineral deposits are commonly arranged in zones around an igneous source; for example, around a granite mass, tin minerals may be found closest to the granite (or even within it), copper minerals next, and lead and zinc minerals furthest away. Basic rock masses have rather different minerals associated with them, such as nickel and platinum minerals.

Mineralizing fluids may fill any available cavities in rocks – pore spaces in sediments, gas cavities (**amygdales**) in lavas, for example. Some minerals form concentrations during the early stages of crystallization of igneous rocks; for example, magnetite (an iron oxide), ilmenite (an iron titanium oxide), and chromite (a chromium iron oxide) concentrations are not uncommon in basic masses.

Ore deposits containing sulphides are especially susceptible to attack by the atmosphere and water. When this happens, sulphides are converted to oxides and carbonates (and occasionally even the free metal) and considerable leaching of the ore body occurs. In a copper sulphide vein, for example, all the copper may have been leached from the upper part, leaving a mass of limonite (an iron oxide containing water). Below this may be a mass of copper oxides, carbonates, and even free copper. Below this again is a zone of rich copper sulphides above the unaltered copper ore. These enriched and oxidized ore bodies are especially valuable.

Ore deposits associated with sedimentary rocks

The processes of sedimentation may concentrate several important minerals. A river carrying sediment may carry away the lighter minerals such as quartz, leaving concentrations of the denser minerals behind. In this way, valuable deposits of gold, diamonds (and other gems), cassiterite, and platinum have been formed. Such deposits are termed **placers**. Ancient ones, called **fossil placers**, are sometimes found in rocks many millions of years old.

In certain circumstances, high concentrations of iron compounds (carbonates, oxides, silicates) form during sedimentary processes and are extensively exploited. Some beds appear to have formed directly from iron-rich sea water, while others appear to be due to the action of iron-rich waters on pre-existing rocks, mainly limestones.

A special group of minerals, mainly chlorides and sulphates, develop when large bodies of salt water evaporate. These are economically important because they are the main source of sodium, potassium, and magnesium compounds.

ROCK-FORMING MINERALS

Most rocks are made up of silicates with minor amounts of oxides. One group of sedimentary rocks, the limestones, consists of carbonates. This account of some rock-forming minerals is not intended as a guide to their identification but as a commentary on some of their more interesting features.

The olivine group

Olivines comprise a group of magnesium iron silicates in which the silicon-oxygen tetrahedra exist as separate units. This structure is nearly uniform in all directions so that there are no planes of weakness and no cleavage. This compact structure makes the mineral hard (6·5 on Mohs' scale) and of moderate density (3·5). Olivine is commonly shades of green, ranging to brown and black in the more iron-rich varieties. Olivine occurs rarely in clear, transparent orthorhombic crystals, and can then be cut as the gemstone, peridot.

Olivine is found as a constituent of basic igneous rocks, such as basalt, dolerite, and gabbro, and of ultrabasic ones – peridotite, for example, is nearly pure olivine. Certain marbles (metamorphosed limestones) contain a magnesium-rich variety, forsterite. Olivine is

Top Magnetite-rich layer formed during the early crystallization of an igneous rock. *Bottom* Ore mineral filling spaces between pebbles in a conglomerate, and filling amygdales in a lava.

igneous rock

magnetite ore mineral

country rock

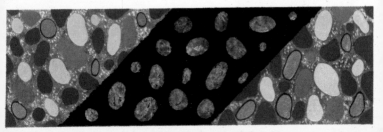

unstable in the presence of water so that it is never found in sediments. Olivine-rich rocks are sometimes altered to the mineral serpentine by reaction with hot waters from the depths.

The pyroxene group

Pyroxenes have a structure consisting of silicon-oxygen tetrahedra united in chains, which are packed together in parallel fashion. The way they are packed develops two prismatic cleavages which are almost at right angles to one another. This is an important test when identifying this group. Pyroxenes are generally hard (5 to 6) and

olivine crystal

granular mass of olivine

augite crystal

hornblende crystal

fibrous mass of actinolite

veins of asbestos in serpentine

moderately dense (3 to 3·6). They are generally brown, green, or black. Good crystals are not common. Simple, iron-magnesium pyroxenes, as typified by hypersthene, are orthorhombic, while types containing calcium, such as augite and diopside, or sodium, such as aegirine, are monoclinic.

Pyroxenes occur in basic and ultrabasic igneous rocks, and sometimes in intermediate types (diorite, andesite). Several types are found in metamorphic rocks and sometimes form the bulk of the rock. They are rare in sediments. Pyroxenes are frequently found altered to serpentine or chlorite.

The amphibole group

Amphiboles have a double chain or 'ribbon' structure of silicon-oxygen tetrahedra. The packing of these 'ribbons' produces two prismatic cleavages with an angle of approximately 120 degrees between them. This is the simplest way of distinguishing amphiboles from pyroxenes, which may otherwise be difficult. Hardness is about 6, density 2·9 to 3·5. Amphiboles are white, green, brown, or black. Several fibrous varieties occur and the term, asbestos, is applied to some of them. (Note: 'asbestos' is also used for other fibrous silicates.) All the common amphiboles are monoclinic, the commonest being hornblende which is found in a wide range of igneous and metamorphic rocks. The green, needle-like actinolite is found mainly in metamorphosed limestones and basic or ultrabasic igneous rocks. Amphiboles often show alteration to chlorite.

The mica group

Micas have a sheet structure of silicon-oxygen tetrahedra in which the links between the sheets are weak, giving an excellent cleavage into thin, elastic leaves. This is their most characteristic property. Two common types can be recognized: a white mica, muscovite, and a black or dark brown one, biotite. Micas are soft (2 or 3) and moderately dense (2·7 to 3). Crystals (monoclinic) are not common, and usually show a hexagonal cross-section.

Large masses of mica (called **books**) are found in pegmatites, and yield mica for commercial purposes. Micas are valuable for their heat-resisting and electrical-insulating properties. Micas occur in acid and intermediate igneous rocks such as granite, syenite, and trachyte, in many metamorphic rocks (schists and gneisses), and occasionally in sediments (micaceous sandstones).

The feldspar group

The feldspars comprise the commonest and most important group of rock-forming minerals. Their three-dimensional framework structure yields blocky crystals with two cleavages approximately at right angles to one another. They are generally white, cream, pink, green, or grey; less commonly red, dark grey, brown. Hardness is 6, density 2·5 to 2·7.

Two groups are recognized: the alkali feldspars (containing potassium and sodium) of which monoclinic orthoclase and triclinic microcline are the best known; the triclinic plagioclase feldspars which range in composition from a pure sodium type, albite, to a pure calcium type, anorthite, through varieties containing both sodium and calcium. A characteristic of plagioclase is the fine lamellar structure often seen on one cleavage surface due to repeated twinning. Orthoclase crystals often show two distinct units of a simple twin.

Alkali feldspars occur in acid and some intermediate igneous rocks such as granites, syenites, rhyolites, and trachytes while plagioclase is most commonly found in basic and intermediate rocks such as gabbro, diorite, basalt, and andesite. Metamorphic rocks frequently contain alkali feldspar or sodium-rich plagioclase. Feldspar decomposes in wet conditions to kaolin or china clay, so that it is only found in sediments which have formed in dry conditions in deserts, for example, or during very rapid deposition and burial such as during sheet floods.

Feldspar is used as an abrasive and in glazes for china. Kaolin is used in making porcelain, glossy paper, and as a filler in paints, rubber, and so on.

The feldspathoid group

A minor group, feldspathoids are found only in certain igneous rocks. The two most important are nepheline (hexagonal) containing sodium, and leucite (cubic) containing potassium. Both have framework structures, a poor cleavage, and are white or pale grey. Leucite usually forms good crystals, while nepheline is massive with a distinctly greasy appearance. Both are hardness 5·5 to 6 and density 2·5.

Feldspathoids occur in certain uncommon basic and intermediate rocks which contain excess amounts of sodium or potassium and a deficiency of silica. The blue feldspathoids, lapis lazuli and sodalite, are used decoratively.

The quartz group

Quartz and its varieties are among the commonest minerals known, occurring widely in nature. Quartz consists of silicon and oxygen linked together in a continuous framework of atoms, chemically known as silica. No cleavage is developed but the mineral breaks with a conchoidal fracture. Quartz is hard (7), with a density of 2·6. Many colour varieties exist; for example: clear and transparent, known as rock crystal; violet quartz is amethyst; pink is rose quartz; smoky yellow or brown is cairngorm; black is morion; opaque white is milky quartz; the colours are due to the presence of traces of impurity.

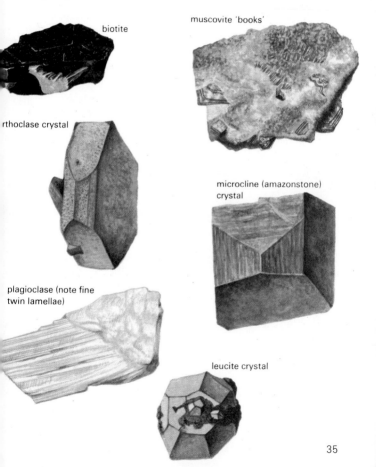

biotite

muscovite 'books'

rthoclase crystal

microcline (amazonstone) crystal

plagioclase (note fine twin lamellae)

leucite crystal

35

Hexagonal crystals of quartz are fairly common and occasionally they reach a metre or more in length. Rock crystal is important in optical work and in certain electronic applications. Cryptocrystalline silica is known as chalcedony and is a very fine-grained, sometimes acicular, aggregate occurring in a wide range of colours. The purer forms are yellowish or greyish with a rather greasy appearance, often in nodules, branching or as globular aggregates. Red varieties are known as jasper or carnelian; prase is green; bloodstone is dark green with red spots. Banded varieties are termed agate. Flint (found in chalk) and chert (in older limestones) are black, compact varieties, very resistant, surviving as pebbles on beaches and in younger rocks. Opal is amorphous silica, containing water. Gem quality opal exhibits a play of colours but other varieties exist, for example, hot spring deposits of siliceous sinter.

Quartz is found in acid igneous rocks and associated mineral veins. Small quartz veins are widespread, sometimes as a result of igneous activity, sometimes due to solution of material from surrounding rocks with deposition in an adjacent crack. Quartz is resistant so that it is common in sediments (sandstones). It is also found in many metamorphic rocks, notably quartzites (metamorphosed sandstones). Chalcedonic silica occurs in sediments, as a cavity filling in lavas, and in veins. Most opal has been deposited from

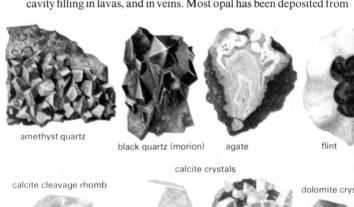

amethyst quartz

black quartz (morion)   agate   flint

calcite crystals

calcite cleavage rhomb   dolomite crys

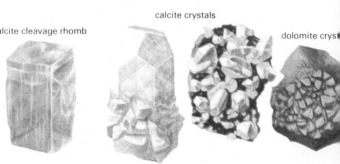

solution, either in sediments, cavities in lavas, around hot springs, or replacing the soft tissues of plants or animals (for example, wood opal).

The carbonate group

Carbonates are virtually the only rock-forming minerals which are not silicates. The two most important are calcite (calcium carbonate) and dolomite (calcium magnesium carbonate). Both are trigonal and show an excellent cleavage into regular rhombohedra. The hardness is 3 for calcite and 3·5 to 4·0 for dolomite and the densities are 2·7 and 2·8 respectively. Calcite occurs in characteristic crystal forms such as rhombohedra, nail-head crystals (hexagonal prism terminated in three flat faces) and dog-tooth crystals (prism terminated in six steep faces, giving a pointed crystal). Crystals of dolomite are usually simple rhombohedra, commonly with curved faces, but are rarer.

Many animals build skeletons of calcite – sea-urchins, corals, crabs and 'shellfish' such as oysters, scallops, whelks, and so on. Such skeletons may constitute much of the calcite of limestones. Crystals of calcite are commonly found in veins and lining joints and cavities in limestones, in which they are deposited from solution. Stalactites and stalagmites are made of calcite. Calcite also occurs in mineral veins, associated with igneous rocks, and filling cavities in basalts. Marbles are metamorphosed limestones and are sometimes almost pure calcite. Optically clear cleavage pieces of calcite have the property of double refraction (a spot looked at through the crystal appears double) and are used in certain optical instruments.

Dolomite is less common than calcite. It often occurs as a replacement of calcite in dolomitic limestones, where it is probably produced by reaction with magnesium-containing solutions. Some dolomite is deposited during the early stages of the evaporation of a mass of sea water. Dolomite is less common as a vein mineral. Massive dolomite is used in refractory bricks in furnaces.

All carbonates react with acid, effervescing as carbon dioxide is evolved.

Other silicate minerals

Two rock-forming minerals of particular interest are tourmaline and topaz. Tourmaline is built up of rings of six silicon-oxygen tetrahedra, together with triangular units consisting of boron and three oxygen atoms. This results in trigonal crystals having a characteristic triangular cross-section. Tourmaline is a very hard mineral (7·5), has no cleavage, and a density of 3 to 3·2. The colour range of tourmaline is enormous – colourless, shades of pink, red, yellow, orange, blue, violet, green, brown, and black are all known, and because of their occurrence as material of gem quality, they have frequently been cut and offered as substitutes for more valuable gems, sometimes under names like Brazilian emerald and Brazilian sapphire. Sometimes crystals of tourmaline show two or more distinct colours due to changes in the elements available during crystal growth; for example, the centre of a crystal may be red, the outer part green, or one end of a crystal may be yellow, the other red. Radiating masses of black needle-like crystals sometimes occur, known to miners as schorl.

Tourmaline mainly occurs in granites, especially in pegmatites, in which giant crystals have been found. Tourmaline in granites may replace earlier-formed minerals such as biotite, perhaps due to reaction with boron-containing fluids left after the crystallization of the main mass of igneous rock. These residual fluids could also account for tourmaline in pegmatites and quartz-tourmaline veins. Tourmaline also occurs as resistant grains in sandstones.

Topaz is an aluminium silicate, containing hydroxyl and fluorine. It is orthorhombic, very hard (8), and dense (3·5). It has a good basal cleavage. Topaz may be colourless, bluish, yellow, orange or pink; much pink topaz sold by jewellers is artificially produced by heating yellow material. Gem quality topaz is not common. Topaz develops in granitic rocks which have been subjected to the action of fluorine-rich vapours and solutions and in association with tourmaline.

tourmaline crystals

topaz crystals

Some metamorphic minerals

Some minerals are found almost exclusively in metamorphic rocks; for example, the garnets, the aluminium silicates, and some calcium silicates.

The garnets are cubic, with a structure of separate silicon-oxygen tetrahedra. They are all hard (6·5 to 7·5), of high density (3·5 to 4·3), and lack cleavage. Two main groups are recognized: those containing essential calcium (grossularite) and those containing essential magnesium and iron (almandine and pyrope). Although we normally think of garnets as red, they are commonly green, brown, or black. Fist-sized crystals have been recorded, and specimens 2 or 3 centimetres in diameter are not uncommon.

Grossularite is mainly found in metamorphosed limestones and sometimes in metamorphosed basic igneous rocks. Almandine and pyrope are common in many kinds of intensely metamorphosed sedimentary rocks – schist, gneisses, hornfels. Pyrope is found in rocks which have been subjected to high pressures. Garnets are hard enough to occur as grains in sandstones or even pebbles in gravels. Gem material, usually almandine or pyrope, comes mainly from stream gravels.

The aluminium silicates comprise the three minerals andalusite, kyanite, and sillimanite which differ in structure and properties even though they have a similar composition.

Andalusite occurs in square, prismatic crystals (it is orthorhombic but looks tetragonal). The variety, chiastolite, shows a black cross of inclusions on an end section in certain kinds of slate. Andalusite is usually greyish or pinkish and often coated with white mica. The hardness is 7·5, density 3·2, and it has a poor prismatic cleavage. It is commonly found in thermally metamorphosed, clayey, sedimentary rocks.

Sillimanite is also orthorhombic, but is commonly fibrous, in wisps. It is greyish or brownish, with hardness 6·5, and density 3·2. Sillimanite is a high-temperature metamorphic mineral developing in clayey rocks, especially if they are subjected to pressure.

Kyanite is triclinic and commonly occurs in bladed crystals in various shades of white, blue, and green, sometimes patchily arranged in a single crystal. Three cleavages are developed, roughly at right angles to one another; one is much better than the others. Kyanite shows varying hardness in different directions, ranging from 4 to 7, according to which crystal face is being scratched. Density is 3·6. Kyanite is found in strongly metamorphosed, clayey rocks, sometimes in quartz veins and 'clots' associated with such metamorphism. Grains of kyanite sometimes occur in sandstones.

The simplest calcium silicate is wollastonite, which is triclinic with a structure based on chains of silicon-oxygen tetrahedra – similar to, but distinct from, the chains in the pyroxene structure. Wollastonite is commonly found as white masses of needle-like crystals, often in radiating aggregates. Hardness 5, density 2·8, it has one good cleavage not easily seen in most specimens. Wollastonite is found in thermally metamorphosed limestones, often in association with calcium garnets.

VEIN MINERALS

Most vein minerals are non-silicates although they are often associated with silicates. Some, such as magnetite, occur as rock formers.

Native elements

Few elements occur native (that is, in a free state) in the Earth's crust. Of the metals, only copper, silver, gold, and the platinum group occur in workable amounts, and gold and platinum are' so rare that almost any amount is workable! Native copper and silver sometimes occur in the upper, weathered parts of sulphide veins. Near Lake Superior in Canada, native copper cements a pebble bed and fills gas cavities in lava flows. Native gold and platinum are best known as grains and nuggets in stream deposits, from which they are recovered by panning. Such deposits are derived from primary sources such as gold-bearing quartz veins and acid igneous rocks, and platinum-bearing, ultrabasic, igneous rocks.

Native non-metals include sulphur, found around volcanic centres, and carbon, in the two forms graphite and diamond. Graphite (hexagonal) is soft (2), black, metallic looking, with a good basal cleavage and contrasts sharply with diamond which is the hardest (10) natural substance, colourless, glassy looking, cleaving into pyramidal forms of cubic type. Graphite is commonly found associated with metamorphic rocks. Some may form from primary carbon in the crust but some arises from metamorphism of organic carbon. Diamonds originate in volcanic explosion-vent breccias of ultrabasic composition, presumably deep in the Earth at extreme temperatures and pressures. They also occur as derived grains ánd pebbles in sediments, the source of most of the diamonds for jewellery and industrial uses.

garnets

andalusite: variety chiastolite

wollastonite   garnet

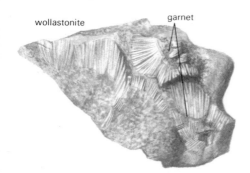

kyanite

Oxides

Among the main elements which occur commonly as oxides are iron, aluminium, tin, and manganese. There are three common iron oxides, all important ores of iron: magnetite, haematite, and 'limonite'. The latter differs from the first two by being hydrated and of variable composition. The properties of the three oxides may be summarized in a table.

	magnetite	**haematite**	**'limonite'**
crystal system	cubic	trigonal	none
colour/lustre	black, metallic	black/metallic, red/earthy	black, brown yellow red/dull or earthy
streak	black	red or red-brown	yellow
hardness	6	6	5
density	5·2	5	3·8

Magnetite and haematite have very poor cleavages. Magnetite is easily recognized because it is magnetic – it is the lodestone of antiquity – and will affect a compass needle or even attract iron objects. Haematite occurs in two distinct forms – black, glistening metallic, crystals, and dark red, kidney-shaped aggregates. It is non-magnetic. 'Limonite' is widespread and usually regarded as a mixture of materials with a variable composition. It is the commonest colouring matter in nature, forming whenever an iron-containing mineral weathers. Almost any brownish stain seen in or on a rock is likely to be 'limonite'. Magnetite is common in many igneous rocks in small amounts, and occasionally in basic rocks in large amounts. It also occurs locally concentrated in beach sands. Haematite occurs in veins, often with quartz, as a replacement in limestones, and by metamorphism of other iron ores.

Aluminium oxide is corundum, second only to diamond in its hardness (9). Corundum is hexagonal, density 4, with no cleavage, but crystals can break into rather rough, tabular fragments. Its colour ranges widely – grey, white, blue, red, greenish, brown, black, and so on. Transparent blue and red varieties are sapphire and ruby respectively. Corundum occurs mainly in metamorphosed, aluminium-rich, silica-poor rocks, and rarely in certain igneous rocks.

Bauxite, the main ore of aluminium, is a mixture of hydrated aluminium oxides with rather ill-defined properties. It may be white, yellow, red, brown, and so on depending upon the amount of iron

copper

graphite

sulphur

haematite crystals

radiating fibrous limonite

granular magnetite

chlorite

magnetite crystals

rock

cassiterite crystals

corundum crystals

oxides present. Small, spherical concretions may be present. Bauxite originates from the weathering of igneous rocks under tropical conditions. Material formed in a similar way but composed of iron oxides, is called laterite.

Cassiterite, the main ore of tin, is a hard (6·5), dense (7), tetragonal mineral. It is generally black, rather metallic, with no cleavage. It occurs in granites, pegmatites, and quartz veins in or close to granites. Much cassiterite is worked from local concentrations in river gravels and sands.

Sulphides

Sulphide minerals are the source of many elements in common use. The four common sulphides described are common enough to be found on many old mine tip heaps. The four are pyrite (iron sulphide); chalcopyrite (copper iron sulphide); galena (lead sulphide); blende or sphalerite (zinc sulphide).

	pyrite	chalcopyrite	galena	blende
crystal system	cubic	tetragonal	cubic	cubic
colour	pale yellow	deep yellow	lead grey	brown, black
lustre	metallic	metallic	metallic	resinous
streak	brownish black	greenish black	grey-black	brownish
hardness	6	4	2·5	4
density	5	4·2	7·5	4
cleavage	nil	nil	three directions into cubes	four directions, but often not easily seen

All these minerals occur in veins associated with igneous rocks, commonly with minerals such as barytes, fluorspar, calcite, and quartz.

Sulphides also occur as replacements, especially of limestones. Chalcopyrite, galena, and blende are important ore minerals, but pyrite is more commonly worked for the sulphur it contains to manufacture sulphuric acid. Pyrite also occurs in sedimentary rocks where it is widespread in certain black shales, cementing some sandstones, and with accumulations of bones and teeth of animals. This type of occurrence suggests that the environment of deposition was almost completely deficient in oxygen. Pyrite occurs as radiating nodules in chalk, but their origin is not fully understood.

Distinguishing between pyrite and chalcopyrite is important, because they are easily confused. Pyrite is harder, occurs more commonly as crystals, and has a palish, brassy colour, as opposed to the distinctly golden colour of chalcopyrite, which is commonly tarnished to brilliant blues and greens. Both minerals have been called fool's gold.

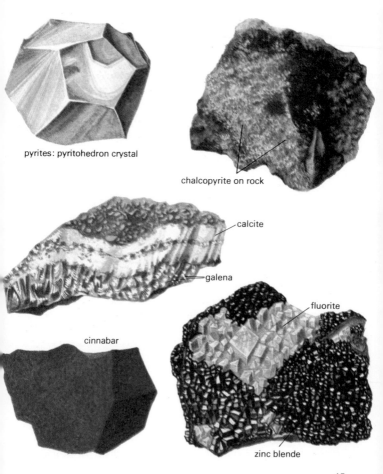

pyrites: pyritohedron crystal

chalcopyrite on rock

calcite

galena

cinnabar

fluorite

zinc blende

Halides

The only common minerals containing elements of the halogen family (fluorine, chlorine, bromine, iodine) are fluorite (calcium fluoride) and rock salt or halite, (sodium chloride).

Fluorite (also called fluorspar) is a common vein mineral which also occurs as a replacement in limestones. It occurs in glassy-looking cubic crystals which often display a good octahedral cleavage resulting in the corners of the cube being truncated. Fluorspar occurs in a wide range of colours – transparent, white, yellow, orange, brown, blue, violet, purple, green, pink – and sometimes shows two or more colours in one crystal. The hardness of fluorspar is 4 and density 3·1. The variety showing blue/violet and white banding is the well-known blue john, much prized for ornamental work (p 23).

Halite or rock salt is one of the very few minerals soluble in water which has a distinctive taste. It occurs in beds and is thought to have formed by the evaporation of large bodies of sea water. It is commonly associated with other minerals believed to have formed in this way, notably the calcium sulphates, gypsum and anhydrite (see later) and dolomite (p 36). Rock salt occurs in cubic crystals of hardness 2·5; density 2·2, and it cleaves into cubes. It is often clear and glassy in appearance but may be tinted various shades of pink, red, and brown.

Sulphates, carbonates, and phosphates

Two of the commonest sulphates are barytes and gypsum.

Barytes is a common vein mineral, frequently occurring in rather glassy-looking orthorhombic crystals. It is noticeably heavy (density 4·5) which distinguishes it from many other, comparable minerals. Its hardness is 3·5 and it has three good cleavages, yielding rather matchbox-shaped cleavage fragments. Barytes occurs in a range of colours – transparent, white, pink, red, bluish, greenish, brown. Crystals tend to be rather platy and often occur in sheaves and clusters. Parallel and radiating fibrous forms also occur. Barytes may sometimes be formed by sedimentary processes.

Gypsum is calcium sulphate containing water. It is characteristically soft (hardness 2) and is usually a white or pink colour. Transparent, monoclinic crystals are not uncommon and are frequently twinned. Its density is 2·3 and it has one excellent cleavage which allows crystals to be split into thin sheets. Gypsum also occurs in granular masses (alabaster), fibrous forms, and rosettes of crystals. Gypsum is formed in two main ways: by evaporation of bodies of sea water when gypsum forms before salt and by the reaction of sulphuric acid, produced by the oxidation of pyrite, with calcite in clays. This often develops individual crystals which sometimes encrust shells. Gypsum is the source of plaster of Paris and is a raw material in the manufacture of sulphuric acid and cement.

In addition to the rock-forming carbonates already described, several others occur mostly as vein minerals. Siderite (iron carbonate) is important. This is harder than calcite (4) and rather denser (3·8), but its cleavage into rhombs and reaction with acid (preferably warm) are similar. It is generally a rather pale colour but rapidly oxidizes to a brownish colour. Siderite occurs in veins often with galena and blende and in sedimentary layers called ironstones.

Two copper carbonates are known: malachite which is green;

malachite

azurite

apatite crystal

gypsum crystal

barytes crystal

and azurite which is blue. Both have a hardness of about 4 and a density in the range 3·7 to 4. They are commonly found together in copper deposits which have been subjected to the action of water and the atmosphere (the oxidized zone). Very often, the recognition of malachite and azurite is the first hint of the presence of copper minerals.

The only common phosphate is apatite, a calcium phosphate containing chlorine and/or fluorine. It is hexagonal, its hardness is 5, density 3·2, and it has no obvious cleavage. Apatite occurs in a wide range of colours, green being common, but shades of red, brown, yellow, and violet are known. It occurs in many igneous rocks and in some metamorphosed limestones. It is also a constituent of bones and teeth, which may contribute to sediments. Apatite is an important source of phosphates for fertilizers.

EXPLORATION FOR MINERALS

The ever-increasing demand for raw materials means that geologists are continually striving to devise new ways of finding valuable mineral deposits, especially now that the obvious ones have long since been discovered and exploited. Hidden ore bodies are often under enormous thicknesses of soil covering the bedrock. The geologist can now use two valuable tools which the old-time mineral prospector did not have available – geochemistry and geophysics.

Geochemistry

When an ore body weathers, traces of the elements present pass into the soil from where they may be taken up by plants or washed into streams. Modern chemical analysis enables the geologist to detect minute traces of the elements concerned so that by collecting samples of soil and stream sediments and testing them he or she can detect a zone where there is an unusual concentration of one or more elements. A simple traverse is shown in the diagram. Soil samples taken from over the vein yield high values of the element; this is termed a **positive anomaly**. If stream sediments are analysed, the ones yielding

Top Geochemical survey. Samples are collected at numbered points. The amount of element at each point is plotted on a graph as shown. The peak at site 4 corresponds to the vein.
Bottom Magnetic survey. Readings of local magnetic field strength are taken at numbered points and plotted on a graph as shown. Peak at sites 3 and 4 corresponds to ore body.

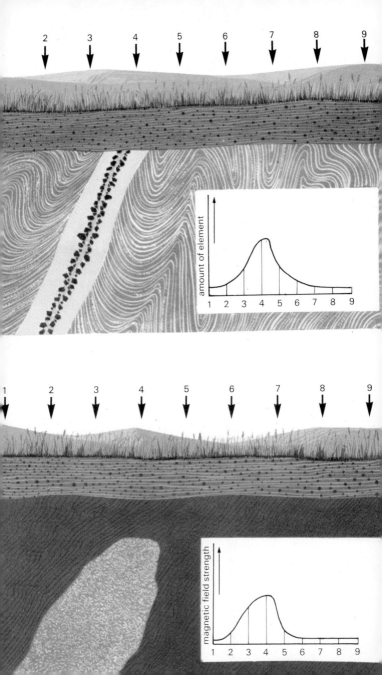

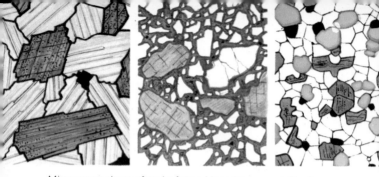

Microscope views of rocks formed by primary crystallization, deposition of fragments, and recrystallization. (See pages 52/3.)

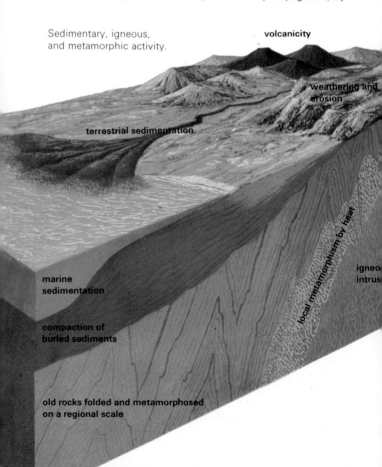

Sedimentary, igneous, and metamorphic activity.

volcanicity

weathering and erosion

terrestrial sedimentation

marine sedimentation

compaction of buried sediments

old rocks folded and metamorphosed on a regional scale

local metamorphism by heat

igneo intrus

the highest values may be followed up for more detailed investigation, and thus a quick survey of a large area may be achieved.

Geophysics

If an ore body has some special physical peculiarity such as an exceptionally high or low density, or it is magnetic, it may be possible to detect its effect on the local, normal values for these properties of the Earth's crust. For example, a body of magnetite can easily be detected by its effect on the Earth's magnetic field (a positive magnetic anomaly) even if it is buried to a considerable depth. Very sensitive magnetic detectors can locate minute variations of the magnetic field, which may yield significant geological information. Rapid regional magnetic surveys can be made by towing instruments in a pod behind an aircraft.

THE STUDY OF ROCKS

Rock-forming processes

'Rock' is an oddly difficult term to define, but *an aggregate of mineral grains* is adequate for most purposes and covers an unconsolidated sand as well as a rigid granite. The study of rocks is the province of the *petrologist*, who usually begins an account of his or her subject by classifying rocks into igneous, sedimentary, and metamorphic (p 10). Before describing these groups, however, it is as well to remark that there are three fundamental processes of rock formation: primary crystallization from a melt or solution, settling out of solids from a suspension, and solid recrystallization. Each is dominant within one of the rock groups but is not confined to it. While primary crystallization is especially characteristic of igneous rock generation, it is also a sedimentary process by which many limestones are formed. And while the settling out of suspensions is of the greatest importance in sedimentary geology, it also plays a part in the igneous field as, for example, in the deposition of volcanic ashes. Metamorphic rocks almost exclusively result from the recrystallization of older rocks but recrystallization also affects recently deposited sediments and cooling igneous rocks.

The imprint of a rock's mode of formation is often retained within it by characteristic features of grain size, grain shape, and grain boundaries, properties collectively referred to as **texture**. Such distinctions can be difficult, however, particularly between primarily crystallized and recrystallized rocks.

Igneous, sedimentary, and metamorphic environments

Igneous, sedimentary, and metamorphic rocks are the products of different environments, because sediments form only at the surface of the Earth whereas metamorphism and igneous activity are mainly internal processes.

The rock melts, or **magmas**, which later crystallize into igneous rocks, originate at depths of 25 to 200 kilometres and temperatures of 1000 °C or more. Moving upwards from their source, the magmas may cool and crystallize as igneous **intrusions** before reaching the surface or they may reach it and erupt as **volcanoes**. With the exception of a few volcanic events the production of igneous rocks takes place above 500 °C. Consequently, their mineral constituents are those stable at high temperatures.

Weathering of rocks exposed to the attack of atmosphere, water, and organisms provides the raw material of sediments. Running water, glaciers, and wind carry away the detritus, and by erosion expose new surfaces to attack. Eventually, the weathering products

The polarizing microscope.

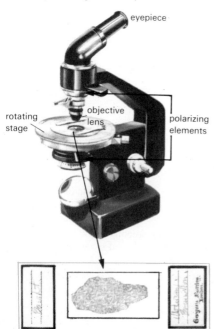

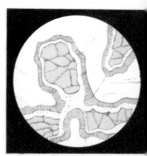

section viewed in ordinary light

mounted **rock thin section**

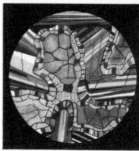

section viewed with polarizing elements in place

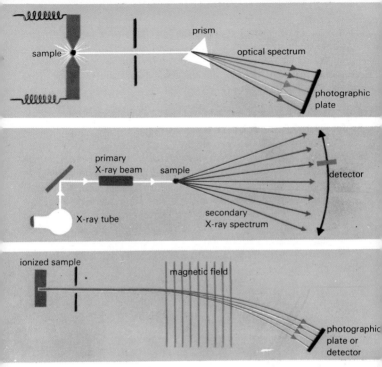

Physical methods of rock analysis. *Top* Emission spectrography.
Centre X-ray fluorescence spectrography. *Bottom* Mass
spectrography.

come to rest in some low-lying land, or in river, lake, or sea. There
they are buried, compacted, and hardened. This whole progression
takes place at near-atmospheric pressures within a temperature
range of −30 °C to 100 °C, and sedimentary minerals must be stable
at these low temperatures and pressures.

Metamorphism is the response of solid rocks to an environment
markedly different from that in which they originated, and may be
triggered off by deep burial or by the rise in temperature near to an
igneous intrusion. The mineralogy of a sediment clearly becomes
inappropriate if subjected to the temperatures and pressures pre-
vailing at 20 kilometres below the surface, and placed in this situation the sediment will react by recrystallizing into minerals stable in
these new conditions. In doing so, it will also change in texture.
Although weathering can be regarded as a type of metamorphism,

it is more convenient to place the lower limits of metamorphism above the temperatures and pressures of surface processes and take as the upper boundary the temperatures and pressures at which rocks melt. So defined, metamorphic temperatures range from 100 °C to 900 °C or so.

Methods used by the petrologist

The first essential of a petrological study is the accurate location, mapping, and description of the rocks in the field. Usually the petrologist next needs to know the mineralogical compositions and textures of the rocks and for this purpose thin sections (see above) are prepared and examined under a **polarizing microscope**. Special features of the microscope include two polarizing elements and a rotatable stage. With only the lower polarizer in place, the rock appears much as it would under an ordinary microscope, whereas with both polarizers in place the rock becomes a mosaic of colours which change in intensity as the stage is rotated. Optical properties

Composition of igneous rocks.

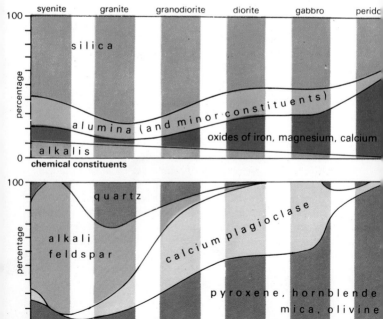

observed in this way help identify the minerals present and reveal many textural features.

If it is necessary to know the chemical composition of the specimens, there are several techniques available ranging from conventional **wet chemical** methods to physical methods based on **spectroscopy** in which a sample is induced to yield a spectrum of measurable radiations which reflect its elemental composition, much as white light can be split by a glass prism into its constituent colours. In **emission spectrography** (p 53), the sample is strongly heated between carbon electrodes to produce a visible spectrum which is recorded on a photographic plate. In **X-ray spectrography**, bombardment by X-rays causes the sample to emit a spectrum of secondary X-rays which is measured by a detector on a moving arm. **Mass spectrography** is a slower but very accurate method in which the sample is broken down into electrically charged ions and a beam of these particles passed through a magnetic field. This produces a spectrum by deflecting the particles to differing extents according to their mass and electric charge.

Zones of magma formation.

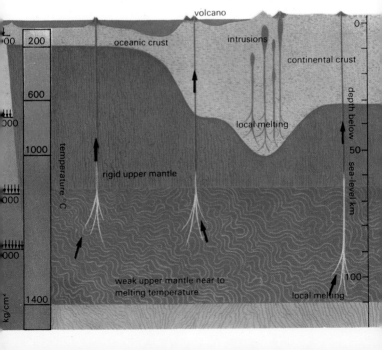

IGNEOUS ROCKS

Magmas and their origin

Lavas pouring down the flanks of a volcano give some insight into the properties of magmas, though both original composition and physical character have already been modified by reactions with near-surface water and the atmosphere. Indirect evidence on the nature of magmas which never reached the surface is obtained by studying igneous intrusions and their metamorphic effects on the surrounding rocks, and this knowledge has been supplemented by numerous experiments in which rocks have been melted and crystallized at pressures equivalent to depths of up to 100 kilometres or more.

Top Crystallization of a melt of two minerals which do not combine to form intermediate compounds.
Bottom Crystallization of a melt of two minerals which form a continuous series of intermediate compounds.

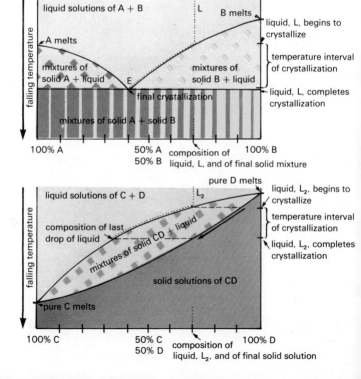

Magmas may be defined as *molten rocks*, though it is rare for them to entirely lack crystalline material. Variations in magma chemistry are reflected in the igneous rocks themselves, the acidic and basic compositions deserving particular attention because of their abundance. Variations in physical properties can be correlated with chemical differences. Thus, temperatures measured on flowing lavas range from less then 700 °C to over 1200 °C, but the lower values are observed on acidic lavas, and experiments show that acidic rocks melt at temperatures as much as 200 °C less than basic rocks. Again, while all magmas are viscous compared with water, an acidic magma may be thousands of times more viscous than a basic melt at the same temperature.

Magmas arrive at the surface with great frequency in volcanically active regions such as the East Indies and Iceland, and at least as much magma must be intruded below the surface, yet earthquake waves which have travelled through the Earth's crust and mantle reveal no permanent liquid layers from which these magmas could have come. We must conclude that magmas arise by local melting of solid rocks at depth. From our knowledge of the rate at which temperature increases with depth, the **geothermal gradient**, and of the melting temperatures of rocks at various pressures, it is possible to locate situations within the Earth's outer layers where the rocks are normally near to their melting points, and it is here that local heating could most readily produce magmas. For acidic rocks, this situation is reached within the continental crust where this is unusually thick. Melting here may be facilitated by water combined in metamorphic minerals, because the presence of water lowers the melting point of rocks.

Melting temperatures of basic rocks are most closely approached at depths between 50 and 200 kilometres, and independent evidence from Earth tremors preceding volcanic eruptions suggests similar levels of magma formation. But these depths lie within the upper mantle which is thought to be of peridotite, a rock different in com-position to basic magma. Experimental petrologists have solved this problem by showing that some peridotites yield a small quota of basic magma if they are melted only partially leaving a solid residue of higher melting point.

Once formed, a magma tends to rise because it is less dense than the rocks which surround it and its progress may be aided by the pressures produced by expansion in the melting zone. Stresses in the Earth's outer layers can also play a part by opening up deep fractures or actively squeezing the magma upwards. By virtue of their low viscosity, basic magmas are able to travel more rapidly and penetrate narrower channelways than acidic magmas.

Solidification of magmas into rocks

A magma is a complex mixture of melted silicates and does not completely crystallize or melt at a fixed temperature. Instead, the process occurs progressively over a temperature change of up to several hundred degrees centigrade. During crystallization, minerals of high melting point appear at an early stage and are followed by minerals of lower melting point, but the precise order varies with the chemical composition of the magma.

The principles governing this behaviour can be appreciated by considering simple melts in which there are only two mineral components. On page 56 A and B are two minerals, such as feldspar and pyroxene, which do not combine to form intermediate compounds. The area at the top of the diagram represents melts of all possible compositions between A and B; the liquid, L, for example, contains 30 per cent A and 70 per cent B. If liquid, L, is cooled, it begins to

Left Crystals settling to the floor of a magma chamber to form layers of varying composition.
Right Granite containing fragments of country rock which have been partly digested by the magma.

crystallize solid B when its temperature reaches the curve joining E with the melting point of pure B. Mineral, B, continues to crystallize as the temperature falls and the composition of the remaining liquid moves towards E as it becomes richer in the A component. At E, solid, A, begins to crystallize and the remaining liquid is used up by the simultaneous crystallization of both minerals which brings the total composition of the solids back to that of the initial liquid. The second diagram shows what happens if the two melted minerals form a continuous series of intermediate compounds, or **solid solutions**, as happens between sodium plagioclase and calcium plagioclase. Liquid, L_2, begins to crystallize at a temperature somewhat below that of pure D and the first crystals have a composition of about 10 per cent C, 90 per cent D, richer in D than the liquid itself. The temperature continues to fall during crystallization and both crystals and liquid become progressively richer in C until the solids finally attain the composition of the initial liquid. At this point, all the liquid is used up, the last drop having the composition 70 per cent C, 30 per cent D.

From these examples, it is evident that mixed silicate melts must always crystallize over a range of temperatures and that the temperature at which a particular mineral starts to crystallize varies with the composition of the melt. Moreover, during crystallization, the compositions of liquids and solids follow different paths and at no point are the two the same. If liquids and crystals somehow become separated during crystallization, the two fractions will form different rocks, one representing the liquid composition and the other the crystal composition at the time of separation. This is an important concept because it follows that the diversity of igneous rocks does not necessarily imply a similar variation in the composition of the initial magmas. If one magma can be split into two or more fractions by the separation of crystals from liquid during crystallization, then there may be only a few 'parental' magma compositions from which all other rocks can be derived.

Acidic and basic igneous rocks are by far the most abundant so that it seems likely that these are the commonest products of melting at depth. Acidic magmas offer little opportunity for fractionation because they crystallize over a small temperature interval but the large temperature intervals of basic magmas make them much more likely parental liquids. If crystals are removed periodically from such a magma, it is evident that the early crystal batches will be composed of high temperature minerals and later batches of lower melting point minerals. One common order in which minerals crystallize from magmas is shown in the diagram and comprises two series of minerals which crystallize independently of each other. If fractionation were to remove crystals in this order from a basic magma, the earliest batches would form ultrabasic rocks, such as peridotite, later batches would give gabbro and diorite, and the final liquid residue would crystallize as a granite.

So far we have not explained why crystals may become separated from their parental magma, but there are several ways in which this might happen, the most important a consequence of the fact that

Textural variation in rocks of the same chemical composition.

granite pegmatite

granit

microgranite

rhyolite

obsidian (glass)

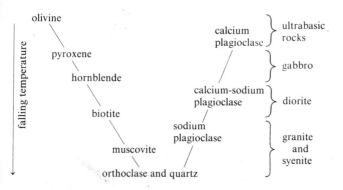

Order of crystallization of minerals from magmas.

most minerals are denser than the magmas from which they crystallize. Depending on their size and on the viscosity of the magma, crystals sink at rates varying from several metres to hundreds of metres per year. A large intrusion can take many thousands of years to crystallize so that there is likely to be ample opportunity for crystals to reach the floor of the magma chamber and accumulate there in layers which in upward sequence record the order in which the minerals were formed. This is exactly what we find in many large intrusions of basic rock.

There is every reason to suppose that crystal fractionation accounts for a good deal of the variation shown by igneous rocks but other processes may also work towards the same end. Mixing of magmas of contrasted composition could result in a whole range of compositions intermediate between them, and at first sight we could imagine almost infinite variation to be possible by the contamination of magmas with solid rocks. This last process can be of only restricted importance, however, because the melting of country rock fragments and reactions between magmas and solid rocks both require considerable quantities of heat, and this can only be supplied from latent heat released by crystallization of magmatic minerals. A magma can incorporate only a limited quantity of solid contaminant before it solidifies.

Textures

Much can be learned about the crystallization history of a rock by examining its texture with the aid of a hand-lens or microscope. Igneous textures are greatly influenced by the rate at which the magma cooled and the order in which the minerals crystallized but other factors such as magma viscosity and movements during crystallization may also leave their mark.

Cooling rate depends upon the size and location of the magma body. Large masses cool more slowly than small ones, and magmas at or near the surface are chilled more quickly than deep intrusions blanketed by warm surroundings. Slow cooling through the crystallization interval allows time for minerals to nucleate around a few

Some igneous textures as seen under the microscope.

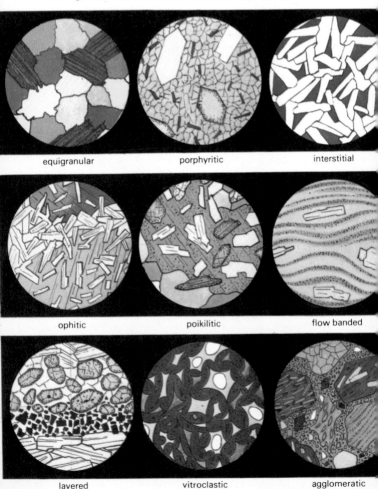

equigranular

porphyritic

interstitial

ophitic

poikilitic

flow banded

layered

vitroclastic

agglomeratic

stable growth centres, giving a coarse grain. Low viscosity also favours coarse crystallization, as does the presence of dissolved water, while the exceptionally large grains found in **pegmatites** are possibly the consequence of crystal growth into bubbles of water enclosed in a water-saturated magma. More rapid loss of heat from a magma can delay crystallization to well below the freezing point, followed by accelerated growth around a large number of nuclei, giving a fine grain. Very fast cooling inhibits crystallization completely so that the magma solidifies to a glass. Many volcanic rocks had already begun to crystallize before reaching the surface and so contain large, slowly-grown crystals in a fine-grained or glassy matrix. This is called **porphyritic** texture, the large crystals being termed **phenocrysts**.

The textural relationships of minerals to each other are mainly governed by the sequence of crystallization. Minerals which begin crystallizing early are able to grow freely and often preserve part of their crystal shape in the final texture whereas later minerals are confined by the surrounding grains and can only fill up the spaces between them. In **poikilitic** textures, these latecomers have grown large enough to enclose small grains of earlier minerals. The name **ophitic** is given to a common variety of this texture in which lath-shaped crystals of plagioclase are enclosed by larger pyroxene grains.

Other textural features reveal that some special mechanism operated during crystallization. Bands differing in mineral composition can be the result of crystals settling to the floor of an intrusion, while flow in a crystallizing magma may cause elongated or platy minerals to lie in a particular direction. Finally, textures akin to those in sedimentary rocks are found in volcanic rocks which were deposited in a fragmented condition.

Classification of igneous rocks

Logical classification of igneous rocks is by no means easy, and numerous schemes have been proposed. For present purposes, it is enough to indicate only the commonest rock types in terms of simple mineralogical and textural features (p 64). Feldspars are the most abundant igneous, rock-forming minerals and they are used to define variations between rocks rich in alkali feldspars and those in which the feldspars are calcic. The presence or absence of quartz provides a supplementary criterion, while abundance of quartz distinguishes the acidic rocks from all others. Compositional groups defined in this way may then be subdivided into coarse-, medium-, and fine-grained varieties.

Volcanoes and volcanic rocks

Volcanic eruptions are impressive testimony to the continued activity of the Earth's interior, and the rock-forming processes they involve are of great importance in the evolution of the crust. A number of large historical eruptions have poured out more than 20 cubic kilometres of rock material on to the Earth's surface, and many prehistoric eruptions were on an even bigger scale, some of them discharging over 100 cubic kilometres of rock. These are exceptional cases, but if the average annual productivity of all the Earth's volcanoes were as little as 4 cubic kilometres, this would be enough to generate a volume of rock equivalent to that of the whole crust in half the time which has elapsed since the Earth was formed.

More than 800 volcanoes are known to have been active over the last few tens of thousands of years, and to this must be added an unknown number of undiscovered submarine volcanoes. No less than four-fifths of recently active volcanoes lie along the margins of the continents which border the Pacific, many of them on arcuate festoons of islands such as those of the East Indies, Japan, and the Aleutian Islands. The Pacific margin is also well known for the frequency and severity of its earthquakes and it contains many of the world's youngest mountain chains. It is perhaps surprising to find that the other main belt of young mountains, extending from the Atlantic to the Pacific through the Alps, Asia Minor, and the Himalayas, has relatively few volcanoes except in the Mediterranean area. Other clusters of young volcanoes are found along the eastern flank of the African continent from the Red Sea southwards as far as Tanzania, following a zone of deep crustal fracturing known as the Rift Valley system.

Classification of igneous rocks.

quartz content	types of feldspar	coarse grained	medium grained	fine grained	genera designati
more than 10 per cent	orthoclase and/or sodic plagioclase	granite	microgranite	rhyolite	acidic
little or none		syenite	microsyenite	trachyte	intermed
	sodi-calcic plagioclase	diorite	microdiorite	andesite	
	calcic plagioclase	gabbro	dolerite	basalt	basic
none	little or no feldspar	peridotite			ultraba

World distribution of active volcanoes.

Less is known about the volcanoes of the ocean basins but their general distribution is well understood. Most of the many hundreds of oceanic islands are composed of volcanic rocks and over forty of them possess active volcanoes. But the islands represent only a small fraction of the total number of oceanic volcanoes. Scattered over the sea floor there are thousands of conical and shield-shaped hills, and nearly all are thought to be extinct volcanoes. Even more voluminous activity occurs along fractures which follow the crests of the **mid-ocean ridges** – the great submarine mountain ranges found in all the oceans. Only in Iceland does a mid-ocean ridge rise above sea-level, and there it forms one of the most volcanically active regions on Earth. Iceland's activity is particularly characterized by the discharge of huge volumes of lava from fissures, and there is every reason to suppose that similar eruptions occur almost as frequently along other sections of the mid-ocean ridges. Taken altogether it seems that volcanicity is an even more important process in the oceans than it is on the continents.

It is often supposed that every volcanic eruption is an exceedingly violent event but, in fact, while all eruptions provide impressive spectacles, they vary enormously in their intensity and in the damage they cause. Quiet effusion of lava typifies many of the eruptions of Hawaii and Iceland while at the other extreme, the 1883 eruption of Krakatao in the East Indies climaxed with explosions heard 5000 kilometres away, threw ash 80 kilometres into the air and caused tidal waves killing 36 000 people. The volcanoes in any one region tend to behave in a similar manner. Those around the Pacific are noted for their explosive nature whereas volcanoes within the ocean basins are for the most part given to quieter activity. Volcanoes emit large quantities of gas as well as magma and the degree of violence with which they erupt is closely linked to viscosity of the magma and the amount of gas evolved. Highly fluid, basic magmas release their contained gases quite easily but the greater viscosity of many intermediate and acidic magmas makes it difficult for the gas bubbles to grow, rise, and escape so that the gas builds up high internal pressures before explosively breaking free.

For those who live near an active volcano, some warning of an impending eruption would be of the greatest help, and methods of predicting such events are a subject of active research. Accurate **tiltmeters** can measure the slow swelling which builds up prior to eruption while delicate **seismographs** detect the weak earth tremors which often accompany the rise of magma towards the surface. It will always be difficult, however, to forecast the likely scale of an imminent eruption.

A volcano's size, shape, and structure are determined by the character of the eruptions which formed it and the volume of material discharged (p 67). Quiet emission of fluid magma builds up a plateau of thin flows if the focus of activity is constantly shifting, but produces a shield-shaped volcano when activity is concentrated at a particular centre. Less commonly, more viscous lava in smaller quantity produces a steeper-sided conical shape. Most conical volcanoes, however, include a considerable amount of fragmental material ejected during explosive activity. This **pyroclastic** debris may build the entire structure but most large cones are **composite** in that they contain tongues of lava interleaved with the fragmental material (p 67). The summit of a big volcano may be marked by a relatively small crater or by a much larger depression several kilometres across known as a **caldera**. Calderas are formed by the subsidence of the whole summit area within a subcircular fracture, an event which is nearly always accompanied by a particularly severe eruption.

The products of volcanicity include gases, liquids, and solids, the gases being largely lost to the atmosphere, the liquids solidifying as lava flows, and the solids depositing as fragmental rocks. Steam is the most abundant volcanic gas but carbon dioxide, carbon monoxide, and compounds of sulphur are also common.

Lava flows vary greatly in character according to their viscosity.

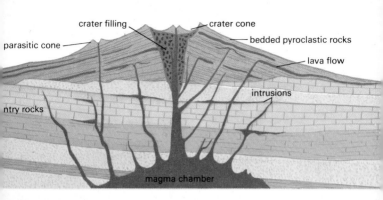

crater filling
crater cone
parasitic cone
bedded pyroclastic rocks
lava flow
intrusions
ntry rocks
magma chamber

Top Structure of a typical stratovolcano.
Centre A lava flow in motion.
Bottom Types of volcano.

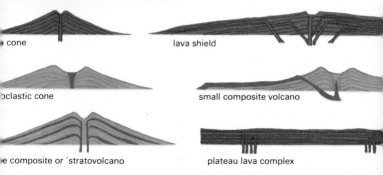

cone

lava shield

oclastic cone

small composite volcano

e composite or 'stratovolcano

plateau lava complex

67

The relatively fluid basic lavas tend to spread out as extensive sheets, some with relatively smooth and ropy-looking surfaces and others covered by cindery blocks and spines. The solidifying front of a flow is often fed by streams of incandescent magma flowing either in open channels or in tunnels beneath a congealed crust. When erupted under water, however, the flow extends forward in small lobes which bud off sack-like bodies of magma producing a distinctive structure to which the name **pillow lava** is given. The rocks composing these various types of flow are mainly basalts and andesites. These dark-coloured, fine-grained rocks commonly carry phenocrysts of olivine, pyroxene, and plagioclase feldspar. Gas vesicles are often present and in the older flows are filled by later-deposited minerals.

The more acidic magmas form short, thick flows or domes, frequently covered with large blocks and seamed by deep crevasses. Owing to their high viscosity, such flows advance only slowly behind a steep front of avalanching debris. The rocks composing them are mainly rhyolites and trachytes normally composed largely of glass though phenocrysts of minerals such as feldspar, quartz, and mica are often present. Glassy rocks are dark in colour but glass is very susceptible to subsequent crystallization in the solid state, so that most older rhyolites and trachytes are pale grey or pink rocks of very fine grain. Vesicles are less numerous than in basic lavas but a strongly banded appearance is very common and is due to the streaking out of compositional inequalities during flow.

Among the rocks generated by explosive ejection from a vent, the coarse-grained **agglomerates** and finer-grained **tuffs** are the chief types. The larger fragments may include pieces torn from the walls of the vent, rounded or contorted **bombs** produced by spattering of still-liquid magma, and highly vesiculated glass known as **pumice**. Most of the smaller fragments are crystals and angular slivers of glass called **shards**. Explosive activity is most vigorous when the magma is intermediate or acidic in composition, so that pyroclastic rocks are most abundant in this range of composition. In one common type of eruption, the fragments are flung from the vent high into the air, the larger pieces following curved trajectories and falling quickly to earth while the finer material remains suspended for some time as clouds of ash and then settles slowly to the ground, the coarser grains arriving first. Consequently, the deposits resulting from these **air falls** are well bedded, and fragment size tends to decrease laterally away from the vent, and vertically within

Common volcanic rocks.

vesicular basalt in hand
specimen and thin section

flow-banded rhyolite in hand
specimen and thin section

andesite

trachyte

agglomerate

bedded tuff

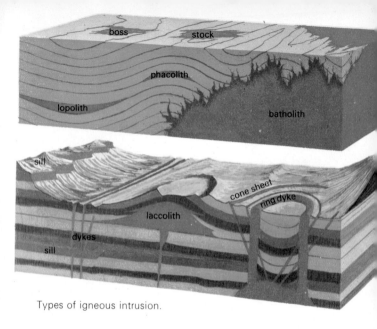

Types of igneous intrusion.

individual beds. Many such layers may be formed in the course of a single eruptive sequence, each explosive pulse sending a fresh shower of material into the air.

Less frequently, a volcano will eject pyroclastic material as an extremely hot, relatively dense mixture of gas and suspended fragments which is not projected into the air but flows downslope with great speed like a thin and turbulent liquid. When these **ash flows** come to rest they are still so hot that the fragments near the base are often compressed and welded together into a compact rock. Such deposits differ from those which result from air fall in that bedding is absent and the fragments are almost completely unsorted in size. Some ash flows are among the most voluminous products of volcanicity. The great rapidity with which they move and their high temperature make ash flow eruptions potentially the most destructive of all.

Top Distribution of sedimentary, igneous, and metamorphic rocks in Europe. Ages in millions of years.
Bottom Major fold belts of Europe. Ages indicate intervals of folding, metamorphism, and igneous activity. Younger sediments cover parts of these belts.

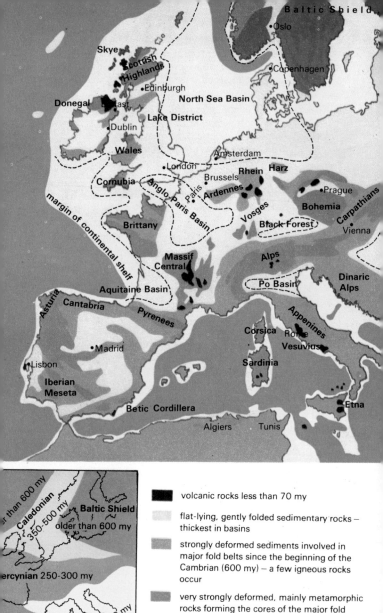

Baltic Shield

Oslo

Skye
Scottish
Highlands
Edinburgh
Copenhagen

Donegal
Belfast
North Sea Basin

Dublin
Lake District

Wales
Amsterdam

London
Rhein
Harz

Cornubia
Brussels
Prague
Ardennes
Paris
Bohemia
Anglo-Paris Basin
Vosges
Carpathians
Brittany
Black Forest
Vienna

margin of continental shelf

Massif
Central
Alps
Dinaric
Alps
Po Basin

Aquitaine Basin
Asturia
Cantabria
Pyrenees
Corsica
Appenines
Rome

Madrid
Vesuvius
Sardinia

Lisbon

Iberian
Meseta
Etna

Betic Cordillera
Algiers
Tunis

[Legend inset map, lower left:]

...r than 600 my
Caledonian
350-500 my
Baltic Shield
older than 600 my

...ercynian 250-300 my

Alpine
12-130 my

[Legend, lower right:]

■ volcanic rocks less than 70 my

flat-lying, gently folded sedimentary rocks –
thickest in basins

strongly deformed sediments involved in
major fold belts since the beginning of the
Cambrian (600 my) – a few igneous rocks
occur

very strongly deformed, mainly metamorphic
rocks forming the cores of the major fold
belts – igneous rocks abundant

rocks older than 600 my, mainly strongly
deformed, metamorphosed, and intruded.

Intrusions and intrusive rocks

Intrusions and volcanoes are often closely associated but some intrusions represent batches of magma which crystallized entirely at depth. Thus, the distribution of intrusive rocks is much the same as that of volcanic rocks. On the continents, intrusions are most numerous in belts of mountain building, both recent and ancient, while other concentrations occur along zones of deep crustal fracturing. Intrusions are undoubtedly very abundant beneath the ocean floors but little is known about them.

Most intrusive rocks are coarser in grain than those cooled more rapidly in a volcanic setting. It is common for the grain size to decrease towards contacts with the surrounding country rocks, a feature known as a **chilled margin**. Granites and gabbros are the commonest intrusive rocks. Granites are usually pale grey or pink in colour, white or pink feldspar and glassy-looking quartz being the chief constituents, accompanied by perhaps 10 or 20 per cent biotite or other coloured minerals. In gabbros, on the other hand, dark-coloured minerals such as pyroxene and olivine make up about half the volume, and plagioclase feldspar composes the rest. The less common peridotites are composed entirely of olivine and pyroxene, are invariably dark in colour, and distinctly heavier than gabbros. Syenites and diorites come next in abundance after granites and gabbros. Syenites resemble granites in their general appearance but contain little or no quartz, while diorites are similar to gabbros but have less coloured minerals.

Intrusions display great diversity in size and shape and some of the commonest types are illustrated on page 70. A distinction is often made between intrusions which are **discordant** in that they cut across the bedding of the country rocks and intrusions which have in-sinuated themselves along the bedding planes and are **concordant** but in either case the magma must make room for itself by displacing the country rocks. Several mechanisms may be involved. If the country rocks are under lateral tension, the magma may open up fractures by exerting only minimal pressure. **Dykes** are often formed in this way, frequently in swarms with thousands of individual members.

Sills and **laccoliths**, which are concordant intrusions, may be formed when the country rocks are under lateral compression, the magma finding it easier to exploit subhorizontal planes of weakness rather than rise higher. Laccoliths updome their roofs so that they

Common intrusive igneous rocks.

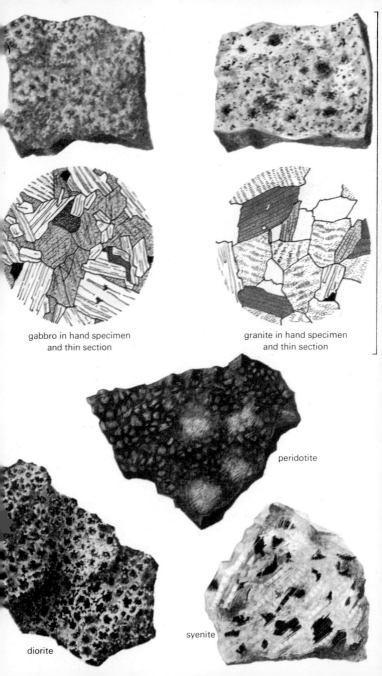

gabbro in hand specimen
and thin section

granite in hand specimen
and thin section

peridotite

diorite

syenite

are found only at shallow depth where the weight of the overlying rocks is relatively small. Many **stocks, bosses,** and **batholiths** have partly made space by forcefully shouldering aside the surrounding rocks which are folded, fractured, and prised apart in the process. Intrusions may also move upwards by wedging open fractures and detaching blocks of country rock which then sink through the magma to deeper levels, a mechanism known as **stoping**.

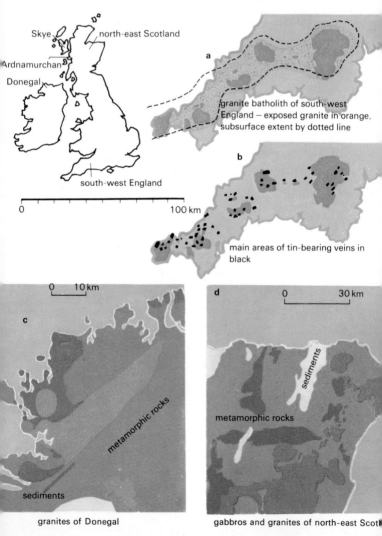

a
granite batholith of south-west England — exposed granite in orange, subsurface extent by dotted line

b
main areas of tin-bearing veins in black

Skye
north-east Scotland
Ardnamurchan
Donegal
south-west England

0   100 km

c
0   10 km
metamorphic rocks
sediments
granites of Donegal

d
0   30 km
metamorphic rocks
sediments
gabbros and granites of north-east Scotl

Batholiths are particularly spectacular examples of the ability of magmas to make space for themselves. Composed principally of granite, they are found in the cores of eroded mountain chains and some are hundreds of kilometres long, but all large batholiths are made up of a number of separate magma injections.

More remarkable for their shape are the ring intrusions – **cone sheets** and **ring dykes** – which are much smaller than batholiths and generally intruded at shallow depth beneath volcanoes. The sub-cylindrical ring dykes are in many cases intruded around the margin of a subsided block which at the surface may be marked by a caldera. Subsidence occurs when pressure is low in the underlying magma chamber whereas high magma pressure can give rise to cone sheets.

A number of British intrusion complexes are shown in the maps here. Those in Skye and Ardnamurchan were intruded about fifty-five million years ago into the cores of large volcanoes and include bosses, ring dykes, and cone sheets, though the last are not shown on the map. Both gabbros and granites are well represented. The Donegal intrusions are much older, about 500 million years, and are typical of the kind of intrusions found in eroded mountain belts. Mainly composed of granite, they include a ring complex, a boss, and a small batholith which is broadly concordant to the country rock structure. The intrusions of north-east Scotland are about 400 million years old and include both granites and gabbros, the latter containing good examples of layering produced by crystal settling in a magma chamber. The granites of south-west England, some 270 million years old, provide another example of intrusion into an ancient mountain belt, and here subsurface connections

Some well-known intrusions in the British Isles. In (c) to (f) granites are shown red and orange, gabbros purple.

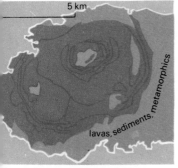

ring complex of Ardnamurchan

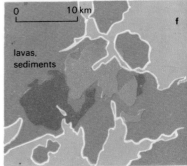

gabbros and granites of Skye

proved by geophysical methods show that the exposed granites are all part of a batholith over 150 kilometres long.

Special interest attaches to south-west England in that it is one of extensive mineralization though most of the veins are now worked out. The close relationship between tin-bearing veins and the granites is clear from the map and the area has also yielded quantities of tungsten, copper, lead, zinc, and other metals. The association of metalliferous veins with intrusions of granite is quite common whereas such mineral veins are only rarely found near intrusions of basic rock. On the other hand, some basic and ultrabasic intrusions contain layers rich in chrome and, more rarely, platinum, while diamonds occur as a minor constituent of an uncommon variety of peridotite known as kimberlite.

SEDIMENTARY ROCKS

Source materials

The ultimate source of most sedimentary material is to be found in the weathering of older rocks on land. Weathering involves the breakdown of rocks by chemical decomposition and mechanical disintegration acting together. Rainwater containing dissolved oxygen and carbon dioxide is capable of decomposing most of silicate minerals by hydration, oxidation, and carbonation, producing mixtures of clay minerals, iron oxides and soluble compounds of

Soil profile.

topsoil highly decomposed rock material plus organic humus

subsoil highly decomposed rock material with little organic content. Often bleached by downward washing of iron and clay

weathered bedrock disintegrated and partly weathered rock material plus iron minerals and clay

bedrock weathered only along fractures

Sedimentary rocks exposed in a sea cliff.

alkalis, magnesium, and calcium. Minerals vary in their suscepti-
bility to this kind of attack, however. Quartz remains unaltered, as
does most muscovite mica, while orthoclase feldspar breaks down
more slowly than plagioclase, and ferromagnesian minerals such as
olivine and pyroxene decompose easily.

Chemical decomposition produces mechanical side effects in that
the volume of the weathering products is greater than that of the
original minerals, and stresses set up by unequal decay of a rock's
components can result in its gradual disintegration without the aid
of other mechanical agents. Normally, however, other forces are
present. Under cold conditions, expansion of freezing water wedges
open joints and grain boundaries, and in hot deserts the large daily
temperature changes may help to disintegrate rocks already weakened
by chemical attack. Organisms also play their part, churning up the
weathering debris and adding humic acids to the percolating rainwater, while roots penetrate into bedrock joints and prise them open.

The products of weathering cover much of the land surface with a
mantle of rock waste in which the chief components are rock fragments, grains of resistant minerals such as quartz, and the newly
formed clay minerals and iron oxides. Wherever the mantle is
stabilized for any length of time, organic activity accelerates the
process of decomposition, adds humus to the inorganic debris, and
produces a soil. Weathering is most active in upland regions, and the
deposits forming the mantle of rock waste are soon removed by
agents of erosion and transported to more permanent resting places
in lowland basins or the sea.

The components of the sandstones, shales, and limestones which dominate most sedimentary rock successions are already present among the products of weathering but it is during transport to the depositional basins that they are sorted out. On land the most important medium of transport is running water but glaciers and wind also carry sediment in regions where running water is scarce. On any hill slope, gravitational mass movements such as soil creep, landslides, and avalanches feed sediments into the drainage system. When sediments reach the sea, tides and currents take over the main transporting role aided by slumping and other mass movements on submarine slopes.

All except glaciers and mass movements are efficient in sorting detritus according to particle size and density, because in moving water or air, the swifter the current the greater the total load which can be carried and the larger the size of particle which can be moved. Wherever the current speed is checked, the coarsest and heaviest particles are the first to be dropped while finer and lighter material is carried onwards. In sorting by size and density, a current inevitably also separates materials of different composition because the majority of large particles are of rock, the intermediate size range is dominated by quartz sand, and the finest material is mostly clay. This mechanical separation is paralleled by chemical separation of the soluble products of weathering as conditions appropriate for their precipitation are encountered, or when organisms abstract some but not others from solution. Potassium is absorbed by clay

Clastic sediments may contain a variety of fragments and cements.

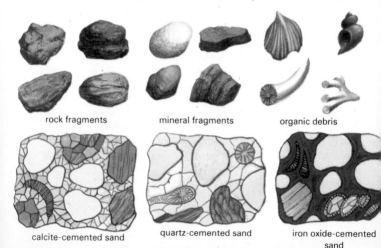

rock fragments

mineral fragments

organic debris

calcite-cemented sand

quartz-cemented sand

iron oxide-cemented sand

Electron microscope view of platy clay mineral particles. Magnified 27 000 times.

minerals or by plants while calcium and magnesium are carried into the sea and there deposited as carbonates. On the other hand, most of the sodium which reaches the sea remains in solution, accounting for the high salt content of sea water.

The transporting mechanisms perform other functions important in determining the character of sediments. Collisions between the larger grains round off the sharp edges of hard particles and fracture softer ones. The moving bed load of a river and the sole of a glacier abrade the rocks below and add to the bulk of detritus. Chemical decomposition continues and increases the stock of clay minerals and soluble compounds. Thus, the components of sedimentary rocks have usually had a long and varied history before they finally arrive at the point of deposition.

Deposition

Sediments are deposited by the settling out of detritus from suspension and by precipitation of minerals from solution, either directly or through the activity of organisms. Sedimentary rocks are broadly divisible into **fragmental** (or **clastic**) and **chemical-organic** classes. More than 99 per cent of all sedimentary rocks belong to three principal rock types of which two, the **shales** and **sandstones**, are composed principally of detritus and the third, **limestone**, includes chemically, organically, and clastically deposited varieties. Shale and sandstone account for more than three-quarters of all sediments, so that it is evident that clastic rocks greatly predominate.

Numerous physical and chemical variables influence depositional processes. The nature of the depositing medium – air, water, or ice –

is obviously of primary importance, and in water, other variables affecting clastic deposition include the concentration of suspended detritus, velocity and constancy of current directions, and turbulence. Chemical sedimentation is controlled by factors such as temperature, concentration of dissolved salts, the relative acidity or alkalinity of the water, and the degree of aeration. These factors also control the nature and amount of biological activity and determine the role of organic deposition.

Structures and textures

The occurrence of sediments in well-defined beds is their most characteristic feature and one which follows from the discontinuous nature of sedimentation processes. The plane which separates one bed from another marks a break in deposition perhaps as long or longer than the time taken to deposit the beds above and below. Some beds are the result of a rapid flush of detritus into the sedimentary basin, others the product of a single season, yet others represent hundreds or thousands of years of slow and undisturbed accumulation. Most beds are laid down on a subhorizontal surface,

Types of bedding and bedding structures.

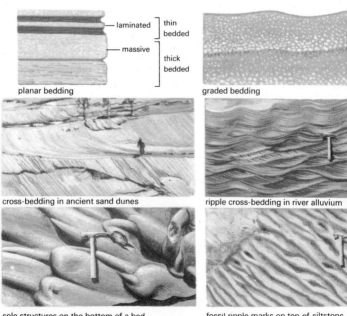

planar bedding

graded bedding

cross-bedding in ancient sand dunes

ripple cross-bedding in river alluvium

sole structures on the bottom of a bed of sandstone

fossil ripple marks on top of siltstone

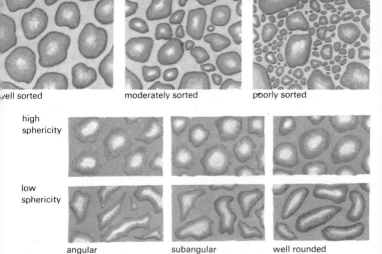

| well sorted | moderately sorted | poorly sorted |

high sphericity		
low sphericity		
angular	subangular	well rounded

Top Sorting.
Bottom Grain shape in fragmental sediments.

though a small depositional slope is normal. Individual beds may be as much as several metres in thickness and few beds are thinner than 1 centimetre.

Useful pointers to the nature of the depositional environment may be found in the internal structure of beds. Of particular importance in clastic sediments are the differences between **plane bedding**, **current bedding**, and **graded bedding**. Plane bedding can be the result of steady settling of detritus in quiet water but often it is formed by deposition from a continuous carpet of moving grains swept along by rapid but smoothly flowing currents. Current bedding, on the other hand, is the product of more turbulent conditions, the inclined surfaces representing the lee side of migrating ripples or dunes or the advancing front of a deltaic sand lobe. Individual current bedded units vary in thickness from a centimetre or so to over 20 metres, the largest and most spectacular being found in desert sand dunes.

At least three processes are capable of producing graded beds. Settling out of a single influx of mixed-size sediment is one such mechanism, as when a shower of volcanic ash is deposited into a lake. A decelerating current can also produce grading because the coarser particles are dropped first followed by successively finer fractions as the current slows. Both these mechanisms achieve a good separation of coarse and fine material, but in the most common type of graded bed, the upward decrease in size affects only the coarser grains and a considerable amount of fine material is present throughout. Such rocks are deposited from dense slurries, known as **turbidity currents**, which often originate as sudden mass movements of sediment down submarine slopes. The high density of these slurries allows coarse and fine material to be transported together and come to rest poorly sorted.

Other structures of environmental significance are often to be found on bedding planes. The upper surface of a bed may preserve ripples similar to those seen today in shallow water or on tidal flats, or it may carry polygonal cracks indicative of the drying out of water-laid sediments exposed for some time to the air. Various kinds of ridges or lobes may project down from the lower surface of a bed filling scours, grooves, or channels in the underlying sediment.

Grain size is one of the most fundamental textural features and is used to make a primary subdivision of clastic sediments into coarse- (**rudaceous**), medium- (**arenaceous**), and fine-grained (**argillaceous**). Within these broad groups there are textural contrasts between sediments in which the grains are well sorted and those which include a wide range of particle sizes. Wind-blown sand and boulder clay provide good examples of the two extremes. Looking beyond grain size to the shape of individual grains, we find every variation between grains which are equidimensional and approach the shape of a sphere, and grains which are flattened, elongated, or irregular in form. Moreover, irrespective of their overall shape, grains may possess angular edges and projections or their edges may be more or less smoothly rounded. Consequently, the shape of clastic particles can be described in terms of **roundness** and **sphericity**. Sphericity is largely determined by the initial shape of the particle as it was released on weathering, but rounding is the result of mutual abrasion of grains during transport.

Perfection of rounding and sorting tends to increase the longer the sediment is carried by the transporting medium so that these two features can be regarded as a rough measure of **textural maturity**. The related concept of **mineralogical maturity** uses as its index the progressive elimination of the less stable rock and mineral fragments. Grain arrangement is yet another aspect of texture. In clastic sediments, the larger fragments often determine the framework, and the spaces between them may be occupied in a number of ways. In well-sorted sediments, even the closest packing of grains leaves a large amount of open pore space which is usually filled, after deposition, by cementing material such as calcite. On the other hand, in poorly sorted sediments, most of the space between the larger fragments is occupied by finer detritus. In some poorly sorted rocks, the fine material is dominant and the larger fragments lie isolated within the clayey matrix.

The textures of chemical and organic sediments are very variable. Some resemble the textures of igneous and metamorphic rocks,

Examples of sedimentary rock textures.

fragmental sediments
Well-sorted, rounded, and closely packed. Large pore spaces filled by cementing material. A texture of highly mature sediments.

Closely packed, poorly sorted, poorly rounded fragments. Abundance of fine material has left little space for later cement. Common in moderately mature sediments.

Very poorly sorted, larger fragments loosely packed in a finely fragmental, muddy matrix with little cement. Typical of immature sediments.

chemical sediments
Interlocking, crystalline texture formed by precipitation from highly saline sea water followed by some recrystallization.

Oolitic textures are common in limestones and ironstones. Formed by the precipitation of carbonates or silicate around sand grains rolling free on the sea floor.

organic sediments
Animal and plant skeletal material is here preserved in growing positions, the spaces between being filled by organic debris and cement. Typical of reef limestone.

Little-broken shells approximately in life positions with abundant cementing material.

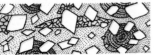

replacement textures
Here, a limestone, mainly calcite ($CaCO_3$) has been partly replaced by rhombs of dolomite [$CaMg(CO_3)_2$] deposited by watery solutions circulating through the rock.

Two places in which sediments are being deposited at the present.
Top Marine sandbanks of carbonate ooliths, Bahamas.
Bottom Dunes of quartz sands, California.

particularly chemical sediments precipitated from saturated saline water, the so-called **evaporites**. These initially possess primary crystalline textures but often later recrystallize in the solid state, in the manner of metamorphic rocks. The oolitic textures of certain chemically deposited limestones and ironstones form another distinctive group. An essential feature of the organic sediments, such as reef limestones and some shell beds, is that a framework of organic skeletal material is preserved more or less in the position of growth. Shell beds, reefs, and the like also produce enormous amounts of sediment by their disintegration and removal elsewhere, but a rock composed of transported organic debris is classified as clastic rather than organic in origin.

Classification

The principles whereby sedimentary rocks may be classified have been referred to on previous pages and are set out in more organized fashion below. In this table the fragmental or clastic sediments are subdivided on grain size whereas the various categories of chemical and organic sediment are based on their chief chemical constituents. Typical members of each of these groups are described in the sections which follow.

FRAGMENTAL			
class	grain sizes mm	fragment types	rock names
rudaceous	— 256 —	boulders	conglomerate and breccia
	— 64 —	cobbles	
	— 2 —	pebbles	
arenaceous	— 1/16 —	sands	sandstone
argillaceous	—1/256—	silts	siltstone
		clays	mudstone, shale

composition	CHEMICAL	ORGANIC
carbonates	oolitic limestone; dolomite	reef and shelly limestones; chalk
siliceous	flint; chert	diatomite; radiolarite
ferruginous	oolitic ironstone; laterite	bog iron ore
aluminous	bauxite	———
phosphatic	phosphorite	guano; bonebeds
saline	rock salt, gypsum, anhydrite	———
carbonaceous	———	peat, lignite, coal

Classification of sedimentary rocks.

Rudaceous fragmental rocks

Rudaceous rocks are by definition coarse in grain but they range from beds of small pebbles or grit up to spectacular accumulations of huge boulders. The particles are so big that they are mostly of rock though spaces between the large fragments, or **phenoclasts**, may be filled with mineral detritus.

Of the two principal rudaceous rock types the **conglomerates** are distinguished by rounded phenoclasts and the **breccias** by angular fragments. Mature conglomerates are often **oligomictic**, that is, most of the phenoclasts are of a single resistant rock type such as quartzite or chert. In such rocks, sorting tends to be good and the phenoclasts close packed, as can be seen in many beach gravels. In contrast, **polymictic** conglomerates contain a variety of rock fragments, rounding is less perfect and sorting often poor, features well displayed by some river gravels.

Whereas conglomerates are always water laid, many breccias form as screes of debris beneath eroding cliffs, and may be oligomictic or polymictic depending on the cliff-forming rocks. Breccias are also deposited by icesheets and valley glaciers as boulder clay and moraine.

Arenaceous fragmental rocks

Quartz is by far the commonest sand-forming mineral, but feldspar and mica contribute significantly to many arenaceous rocks, as do small fragments of fine-grained rocks such as chert and rhyolite while some arenaceous sediments are composed largely of carbonate debris. Moreover, few arenaceous rocks are composed entirely of sand-sized material, an admixture of clayey material being particularly common.

There are three principal types of sandstone, the first being made up almost entirely of quartz grains and known as **orthoquartzite**. This is typically a well-sorted rock with moderately to well-rounded grains cemented by calcite, silica, or iron oxides. Current bedding is common and the deposits as a whole form thin but extensive sheets, often associated with limestones. Evidently, orthoquartzites are mature sediments with a long history of weathering and transport behind them. They are commonly deposited in shallow seas or sandy desert basins.

Common sedimentary rocks (see also pages 89 and 90).

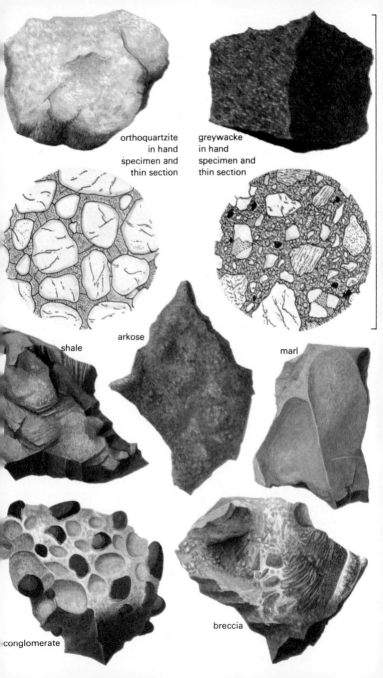

orthoquartzite
in hand
specimen and
thin section

greywacke
in hand
specimen and
thin section

arkose

shale

marl

conglomerate

breccia

Sandstones containing large amounts of feldspar fragments are termed **arkoses**. They are usually rather coarse, often pink in colour, and have cements similar to those of orthoquartzites. Sorting and rounding are only fair, and bedding is often crude, with rough cross bedding in places. Arkoses make up thick, wedge-shaped formations associated with polymictic conglomerates and are usually the product of rapid erosion and deposition in terrestrial basins adjacent to high mountain ranges.

The characteristic feature of **greywacke** sandstones is the high content of argillaceous material, so that these rocks are always poorly sorted and contain little cement. The grains tend to be sub-angular and often include a high proportion of rock fragments and feldspar. Bedding varies from plane to graded and any current bedding is on a small scale. Greywackes are immature both texturally and mineralogically, and like arkoses, appear to have been deposited rapidly. They are mostly of marine origin and were often deposited beyond the continental shelf by turbidity currents.

Argillaceous fragmental rocks

Fineness of grain makes these rocks difficult to study and special techniques are needed to identify many of their constituents. The distinctive compositional feature of many argillaceous sediments is a preponderance of minute, flaky particles of clay minerals crystallized at various stages of weathering, transport, and deposition. Minor constituents include chemically deposited silica and iron oxides, while in **marls** there are appreciable amounts of calcium carbonate. Argillaceous sediments are initially deposited as muds and only during later compaction do they change first into plastic clay and then harden into shale or mudstone. Shales differ from mudstones in that they readily split along planes parallel to the bedding.

Most argillaceous sediments are deposited from water in which currents are weak and there is little turbulence. At the present time they are forming on lake floors, the tidal mudflats of estuaries, offshore from deltas and, most importantly, in the deeper waters beyond the edge of the continental shelf.

Chemically deposited sediments

The iron and aluminium hydroxide rocks known as laterite and bauxite are examples of chemical deposits formed on land by weathering concentration in warm, wet climates. Most chemical deposits are

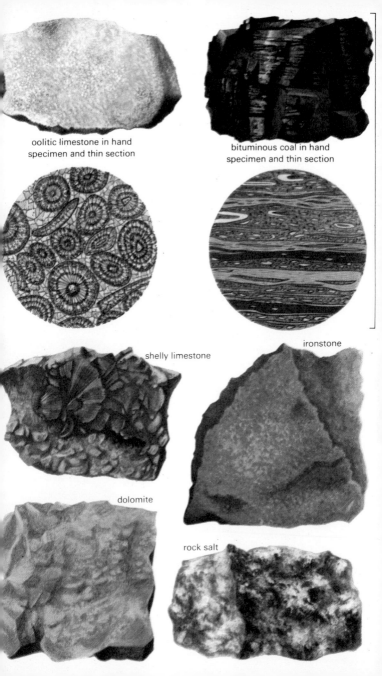

oolitic limestone in hand
specimen and thin section

bituminous coal in hand
specimen and thin section

shelly limestone

ironstone

dolomite

rock salt

formed in water which has become locally oversaturated with dissolved salts, however.

Calcium carbonate precipitates most readily in warm, shallow sea water of moderately high salinity. Structureless carbonate muds may form in quiet water but in more turbulent situations, precipitation centres around detrital fragments rolling to and fro over the sea-floor, building up accumulations of small, subspherical bodies

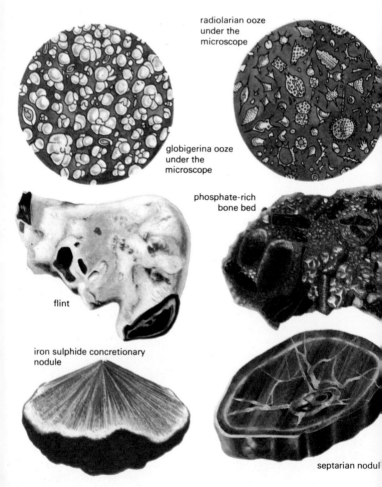

radiolarian ooze under the microscope

globigerina ooze under the microscope

phosphate-rich bone bed

flint

iron sulphide concretionary nodule

septarian nodul

called **oolites**. Many sedimentary ironstones and phosphate rocks are of similar texture.

The commonest inorganic silica rocks are the very fine-grained flints and cherts which often occur as nodules or thin beds in limestone. These may originate by the direct precipitation of colloidal silica brought in by rivers although siliceous organic debris is an alternative source.

The saline deposits, or evaporites, include rock-salt, gypsum, anhydrite, and, more rarely, potassium salts. The largest of these deposits form along seacoasts in hot, dry climates, some of them in bays which act as large evaporating basins and others in salt marshes.

Organically deposited sediments

The material contributed by organisms to sediments is largely in the form of hard, skeletal structures such as shells, spines, and teeth. Organic substances make up a significant proportion of the total bulk of sedimentary rocks but much of this material is in a fragmented state, whereas the term 'organically deposited sediment' should be restricted to organic material which has accumulated more or less undisturbed in the situation in which the organisms lived and died.

Reef limestones provide a good example because they are largely composed of the hard parts of creatures such as corals, molluscs, and algae which congregated in dense colonies on the sea-floor and slowly built up mounds or ridges of skeletal remains on top of which the live organisms dwelt. Reefs grow only in shallow waters and usually in warm climates so that their presence in ancient sediments gives a useful indication of the depositional environment. Limestones composed mainly of shells include both true organic deposits and rocks which are properly regarded as fragmental. Another type of limestone rich in organic remains is the chalky rock originally deposited as calcareous ooze on the ocean floor. The principal constituents are the minute shells of foraminifera which lived as part of the plankton at shallow depths, the shells sinking to the sea-floor after the animals died. Modern examples include the globigerina ooze which is common at depths of around 3600 metres. Of similar origin are the siliceous oozes in which shells of diatoms or radiolaria predominate.

Coal provides an example of an organic rock derived from substances which in normal circumstances decay at the surface. Ordinary household coal and anthracite owe much of their character to changes which occurred after burial. The first and most essential stage is the formation of a peat, an accumulation of plant remains preserved under conditions which discourage bacterial growth. Most of the economically important coals originated as peats on tropical deltas which carried a heavy cover of trees and herbaceous plants, the poorly oxygenated swamp water and rapid accumulation preventing complete decay. After burial, the seams of peat were slowly changed by heat and pressure; water and other volatile constituents were driven off and produced first a dark brown, lignite coal, then at a more advanced stage the familiar, black, bituminous coal. In places, the process went even further and resulted in anthracite, a nearly pure carbon rock of high calorific content.

Where sediments are deposited

Geological processes on land are dominated by weathering and erosion whereas deposition is most important in the sea. This is not to say that there is no deposition on land or erosion in the sea, but it does imply a net transfer of material from land to sea which in the long term is balanced by the periodic uplift of parts of the sea-floor to form new land. Therefore, the overall control of sedimentary processes lies in the structural evolution of the Earth's outer layers, which determines where mountains are raised up and basins subside. In the short term, however, the character of sedimentation is determined by the geographical situation in relation to land forms and climate, and one of the foremost objects of the sedimentary petrologist is to reconstruct the geography of the past using evidence contained in the sediments deposited at the time. Such interpretations rely heavily on comparisons between ancient and modern sediments, using the assumption that processes and environments have not changed too radically with time, at least during the last few hundred million years of Earth history.

Some of the general situations in which sediments are deposited are illustrated in the diagrams opposite. Each of these environments is characterized by an association of related sediments, but considerable variations are possible according to differences in climate and the nature of the sediment being supplied. Thus, recognition of ancient environments is a difficult task which requires an assessment of every aspect of the association – compositional, structural, textural, and organic.

Some important sedimentary environments.

scree

moraine

lake

outwash gravels

lacier margin

alluvial fan gravels

sands of braided, seasonal streams

sand dunes

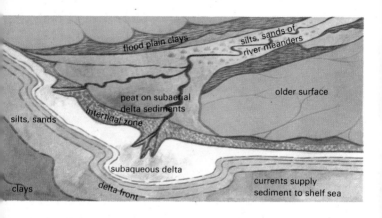

flood plain clays

silts, sands of river meanders

peat on subaerial delta sediments

older surface

silts, sands

intertidal zone

subaqueous delta

clays

delta front

currents supply sediment to shelf sea

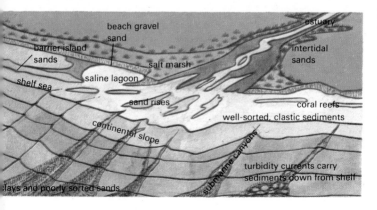

beach gravel sand

estuary

barrier island sands

salt marsh

intertidal sands

shelf sea

saline lagoon

sand rises

coral reefs

well-sorted, clastic sediments

continental slope

submarine canyons

clays and poorly sorted sands

turbidity currents carry sediments down from shelf

Changes following deposition

The conversion of incoherent sands and muds into hard sandstones and shales, and the progressive change of peat into anthracite are examples of extensive modification of sediments after their deposition by processes collectively known as **diagenesis**.

Compaction under the weight of overlying sediment may compress an argillaceous deposit to less than one-half of its original thickness, though sandstone and limestone are only slightly reduced. Water expelled during compaction carries with it dissolved salts which deposit elsewhere as a cement between grains or replace an original constituent by a new one. It is often the case that the calcium carbonate of limestones is partly or wholly replaced by the magnesium-bearing double carbonate, dolomite and, even where not replaced, limestones always undergo some recrystallization.

One curious result of chemical diagenesis is the formation of concretionary bodies by the segregation of some minor constituent of the rock into nodules, lenses, or veins. These structures are particularly common in argillaceous rocks, where they include heavy nodules of radiating, brassy pyrite and dark brown, iron carbonate-rich **ironstone nodules**. Carbonate segregations containing irregularly radiating cracks filled by coarsely crystalline minerals are known as **septarian nodules**.

Organisms also play a part in modifying primary depositional

Sediments disturbed by burrowing organisms. *Left* Siltstone and shale. *Right* Calcareous sandstone with U-shaped animal burrow.

SEDIMENTS	FOSSILS	ENVIRONMENTS
coal	leaves, stems, and spores of trees	swampy forests
seat earth	roots	
mudstone		surface built up to water leve.
sandstone locally cross-bedded with irregularities		deltaic
		influx of river sands
siltstone	plant debris	deltaic lagoons
mudstone and shale		delta front or estuarine
with ironstone nodules	mussels, fish	brackish water
mudstone	marine molluscs	inundation by the sea
coal		swampy forests

A rhythmic sedimentary unit from the Coal Measures and its interpretation.

textures and structures, normally at the earliest stages of diagenesis. Shallow marine sediments are often considerably disturbed by animals which feed on the bottom or burrow into the underlying layers.

Sediments as recorders of Earth history

Although sedimentary rocks make up only a small part of the Earth, they are of unique importance in recording the history of the Earth's surface and the development of life upon it. The evidence is, however, very incomplete. The earliest known sediments were formed more than 3500 million years ago, but it is only for the last 1000 million years or so that we have anything approaching a continuous picture. Organic fossils first become abundant in rocks about 600 million years ago, and by that time already show a high degree of complexity.

The historical interpretation of sediments starts from the obvious fact that in an undisturbed succession of sedimentary beds, the oldest rocks lie at the base and the youngest at the top. If the geologist can find evidence of the original geographical and biological setting of each part of the succession he or she is able to decipher a small piece of geological history. The example shows a sequence of beds which occurs frequently in the Coal Measures of north-western Europe, deposited about 300 million years ago. The upward changes in the nature of the sediments and their fossil content reveal a story which begins with an advance of the sea over an area of tropical forest. The gradual advance of a delta front over the sea-floor builds up sediments to sea-level and allows a heavy growth of vegetation which is eventually terminated by another advance of the sea. These events were repeated many times to give a rhythmically recurring succession of beds in which coals appear at intervals of about 10 to 15 metres. In the sedimentary record there are, however, no continuous successions which span more than a few tens of millions of years because sooner or later deposition is interrupted by uplift and erosion. The evidence is, therefore, scattered and fragmentary.

METAMORPHIC ROCKS

The causes of metamorphism

Metamorphism is the response of solid rocks to an environment markedly different from that in which they originated. This simple idea needs some discussion, however. The mineralogical composition of a sediment is ideally one which is completely stable under the physical and chemical conditions found at the surface of the Earth. Pyroxene and plagioclase feldspar, for instance, are uncommon in sediments because they are susceptible to decomposition in the surface environment, and a sediment which contains them is regarded as mineralogically immature. Put another way, a mineralogically immature sediment is one which has not completely attained a true balance, or equilibrium, with its environment.

The fact that unstable minerals do occur in some sediments indicates that equilibrium is not a condition which rocks reach instantaneously, or even quickly. Indeed, most rocks can persist unaltered for long periods in environments very different from those in which they were formed. If this were not so we would not be able to collect unaltered specimens of igneous and metamorphic rocks at the surface of the Earth because they are here at much lower temperatures and pressures than those which accompanied their formation. Nevertheless, the widespread occurrence of rocks recognizable as having once been sedimentary or igneous though now greatly changed in mineralogy and texture shows that rocks often do readjust themselves to a new situation by recrystallizing in the solid state to a more stable assemblage of minerals.

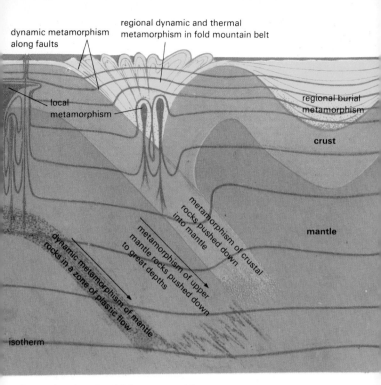

Some situations in which metamorphism occurs.

Evidently, there are factors other than the magnitude of change in temperature and pressure which determine whether or not metamorphic reactions will occur. One of these factors appears to be that rocks respond much more readily to a rise in temperature and pressure than they do when temperature and pressure fall. Secondly, the changed conditions must be maintained for a long enough period to allow time for recrystallization. Thirdly, reactions are often helped along, or catalysed, by the presence of active chemical agents, of which water is one of the most powerful. Finally, rocks under a strong deforming stress react more readily than those which are not.

Where metamorphism occurs
Metamorphic environments can be broadly categorized as either local or regional depending on their scale.

The commonest instances of localized metamorphism are to be found adjacent to igneous intrusions and are termed **contact metamorphism**. In this situation, the rocks are recrystallized in response to a rise in temperature, the pressure remaining more or less unchanged. Basic magmas are emplaced at about 1000 °C to 1200 °C and acidic magmas in the range 700 to 900 °C, so that very high temperatures may be induced in the country rocks immediately against the magma body or enclosed as fragments within it, but temperatures fall rapidly away from the contact because rocks are poor conductors of heat.

An intrusion exposed at the surface reveals the magnitude of its effect on the country rock by the thickness of the envelope of metamorphosed rocks which surrounds it. This envelope, known as the **contact aureole**, is vanishingly small adjacent to most dykes and sills but may be 2 or 3 kilometres wide near a big intrusion. The high temperatures of basic magmas may lead us to suppose that the aureole around a basic intrusion will be wider than that adjacent to an acidic intrusion of comparable size but, in fact, the reverse is more often true. The answer lies in the high water content of many acidic magmas, because water expelled from the crystallizing intrusion carries heat out into the country rocks and also acts as a catalyst to the metamorphic reactions.

Another type of localized metamorphism takes place in the vicinity of some large faults (that is, fractures along which rocks have been displaced), producing **fault breccia** under near-surface conditions, and fine-grained, flinty material at greater depth.

Impact or **shock metamorphism** is a localized effect which has only recently become clearly recognized. It is caused by the impact of meteorites on surface rocks and can produce craters varying from small pits to depressions many kilometres across. Fortunately for us,

Crater made by a meteorite impact, Arizona.

hornfels

schist

Hornfels, formed by local heating near igneous intrusions, and a schist, typical of regional metamorphism.

large impacts are rare; a terrestrial meteorite crater is illustrated opposite and even better examples are found among the craters of the Moon. Apart from cratering, the short-lived but enormously high pressures and temperatures induced by the impact cause severe brecciation and local melting of the immediately adjacent rocks and also generate small amounts of rare high pressure silicate minerals.

The types of metamorphism described so far are caused by localized and geologically brief departures from the normal distribution of temperature or pressure within the Earth's outer layers. The much broader scale which characterizes regional metamorphisms is a consequence of major structural movements affecting large sectors of the crust, often accompanied by equally extensive increases in the flow of heat from the Earth's interior. The term **regional metamorphism** is often used to mean only that kind of metamorphism which affects crustal rocks involved in zones of fold mountain building, but while these **orogenic belts** are certainly the best known of the environments in which metamorphism takes place on a larger scale, equally extensive metamorphism also occurs in several other situations within the crust and mantle.

The more recent of the world's orogenic belts are easily discernible on a map because they include all the major fold mountain ranges (p 112). The belts develop initially as linear series of basins in which thick sediments accumulate. At a later stage, the basins undergo intermittent lateral compression so that both sediments and under-lying rocks are deformed into systems of folds and faults. Finally, the belt is uplifted to form mountains. Metamorphism reaches its height during and immediately following the compressional stage, when the basin floors may be forced down to depths of 20 or 30 kilometres and temperatures can locally reach 800 °C or more. Moreover, metamorphic changes are also promoted by the frequent emplacement of igneous intrusions and the abundance of water contained in the largely marine sediments. Not all the rocks of the orogenic belt are metamorphosed, however; the upper parts of the sedimentary piles and much of the marginal area escape, though deformed by folds and faults.

Regional metamorphism can also take place in the lower levels of exceptionally deep sedimentary basins without accompanying orogenic activity. This **burial metamorphism** may not be common, because some very thick sedimentary successions are completely unmetamorphosed. Where it does occur, the minerals formed are of low temperature type and recrystallization is incomplete.

The igneous rocks which compose the ocean floors have been

Metamorphic zoning on a regional scale in the Scottish Highlands and locally in the aureole of the Skiddaw Granite, Lake District, England.

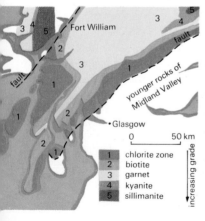

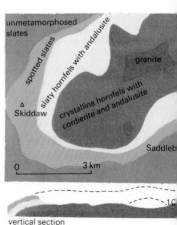

vertical section

subjected to another kind of non-orogenic, regional metamorphism. This takes place beneath mid-ocean ridges, where strong volcanism is associated with high rates of heat flow. Whereas burial metamorphism is the result of depressing sediments into a hotter environment, the ocean ridge metamorphism is caused by a regional temperature rise.

For obvious reasons it is not easy to examine the results of ocean-floor metamorphism 'in the field', and it is even more difficult to assess the effects of metamorphism in the mantle of the Earth. Yet it must be a very common process there. According to modern ideas of Earth structure, great sections of the oceans' crust and upper mantle are pushed down to deep levels beneath active orogenic belts. This must be accompanied by intense shearing along the upper and lower margins of the descending mass, while the increase in temperature and pressure with depth will cause important mineralogical changes, as is predicted from experimental imitation of these conditions. Ultimately, some of these descending rocks must melt.

The nature of metamorphism

The recrystallization of rocks adjacent to an igneous intrusion is a consequence of temperature increase, with little or no change in pressure. Such purely **thermal** effects also occur on a regional scale in the mid-ocean ridges and sometimes in orogenic belts. But much orogenic metamorphism is **dynamothermal** in that it also involves changes in pressure and the application of high directional stresses. Pure **dynamic metamorphism** is exemplified by the crushing and grinding which occur near large faults.

Thermal metamorphism is less effective at changing textures than dynamic or dynamothermal metamorphism and the product is often rather fine grained and may preserve some original features such as bedding. The dynamic element of metamorphism is responsible for imposing on many rocks distinctive new textures characterized by a more or less strong tendency for the grains to grow in particular directions. These differences may be appreciated by comparing a **hornfels**, formed by thermal metamorphism near an intrusion, with a **schist** produced by dynamothermal metamorphism in an orogenic belt.

Metamorphism normally involves little change in the composition of the rock apart from modest losses or gains in volatile constituents such as water and carbon dioxide. Occasionally, however, there are significant compositional changes, perhaps because of an influx of solutions from a nearby intrusion, for example.

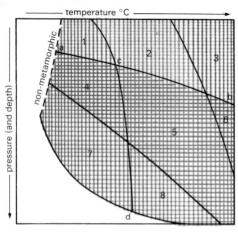

possible zone sequences

x in a contact aureole
y in regionally metamorphosed
   rocks formed in an area of high
   geothermal gradient
z in regionally metamorphosed
   rocks formed in an area of low
   geothermal gradient

Petrogenetic grid.

The mineralogy of a metamorphic rock is determined by its chemical composition and the pressure-temperature conditions of crystallization but, in any one area, the purely physical effects are separately distinguishable as a series of **metamorphic zones** of increasing **grade**. In the vicinity of an igneous intrusion, the grade increases towards the igneous mass and zones can often be mapped by noting the first appearances of distinctive metamorphic minerals as you approach the contact. Regionally metamorphosed rocks are zoned on a larger scale, the grade increasing towards central, more deeply eroded parts of the belt, as, for instance, in the south-west Highlands of Scotland (p 100).

Zones are initially defined in rocks of similar chemical composition, but all the rocks within a given zone must have recrystallized within the same limits of temperature and pressure. Thus, once established, zones allow the correlation of metamorphic changes in rocks of different compositions and define groups of rocks varied in composition but formed under similar physical conditions. Such a group is called a **metamorphic facies**. Some eleven major facies have been recognized in rocks the world over, so that it is now possible to place any local sequence of zones into a much broader picture. Furthermore, much has been done to determine experimentally the pressure-temperature ranges of mineral assemblages characteristic of each facies. The results can be used to construct **petrogenetic grids** in which each box represents the pressure-temperature range of a particular facies. The principle of such a grid can be explained with the help of the diagram. Here, each pure colour shows the range of a single mineral association, yellow, for

Some textures and minor structures of metamorphic rocks.

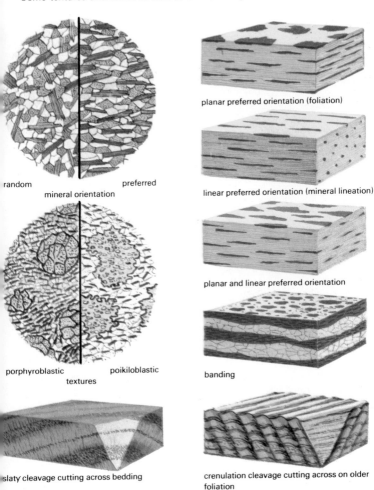

random   preferred
mineral orientation

planar preferred orientation (foliation)

linear preferred orientation (mineral lineation)

planar and linear preferred orientation

porphyroblastic   poikiloblastic
textures

banding

slaty cleavage cutting across bedding

crenulation cleavage cutting across on older foliation

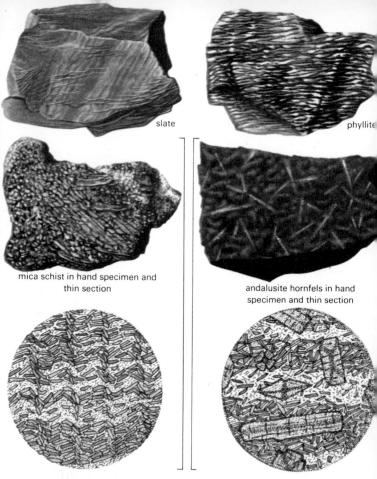

slate

phyllite

mica schist in hand specimen and thin section

andalusite hornfels in hand specimen and thin section

Common metamorphic rocks (see also pages 106 and 109).

instance, representing an association stable over a wide range of temperature but breaking down at pressures higher than a–e–b. The association shown in blue has a wide pressure range but is unstable at temperatures above c–e–d. Thus, a metamorphic facies containing both assemblages is stable only in area 1, bounded by a–e–c. Other facies, numbered 2 to 8 are defined in a similar way. The arrows lettered x, y, and z illustrate some possible sequences of facies zones. Thus, a sequence 1, 2, 3 indicates metamorphism in a high temperature gradient with little pressure change, a situation found near igneous intrusions. The z sequence, on the other hand, owes as much to increasing pressure as to increasing temperature and each zone must have formed at successively greater depth in a region where the temperature gradient was only moderate.

Mineralogy

Many of the minerals found in metamorphic rocks are also common in igneous rocks. The more varied chemical composition of the metamorphic rocks, however, and the wide range of physical conditions in which they can be formed lead to the appearance of minerals which are rare or absent in igneous rocks. Notable examples include some common constituents of metamorphosed argillaceous sediments, such as the aluminium silicates andalusite, kyanite, and sillimanite.

Textures and structures

Regionally metamorphosed rocks are generally coarser than those produced by localized metamorphism, and grains also tend to become larger with increasing grade. Textures superficially similar to those of porphyritic igneous rocks are quite common, with large grains set in a finer matrix, but in this case the large grains, or **porphyroblasts**, are not necessarily the first to grow. Indeed they often form late and may be crowded with enclosed grains of the matrix, a texture known as **poikiloblastic**. Otherwise many rocks consist of more or less equigranular mosaics of minerals.

Rocks formed by dynamic or dynamothermal metamorphism nearly always show some degree of preferred mineral orientation; that is, the constituent grains do not lie entirely at random but tend to grow in particular directions. Preferred orientation is said to be **planar** if expressed as a subparallel arrangement of platy grains and **linear** when it involves elongate grains: frequently both are present. These textures result from mineral growth in rocks in which the stress varies in different directions. Such rocks cleave easily in the direction of planar mineral growth. The strong cleavage of slates, for instance, is due to the parallel arrangement of very fine micaceous grains, usually in a direction which cuts across the original bedding. Sometimes, a second cleavage is superimposed across an earlier slaty cleavage as regularly spaced plans of oriented mica flakes. Examined closely, these planes are seen to lie on the limbs of small folds, or **crenulations**. Folds of various scales are, of course, very common structures in the regionally metamorphosed rocks of orogenic belts. Other structures are inherited from the original rocks, including the sedimentary bedding which often survives contact metamorphism or the lower grades of regional metamorphism. The prominent banding in some high-grade, regionally metamorphosed rocks arises during their recrystallization and should not be termed bedding.

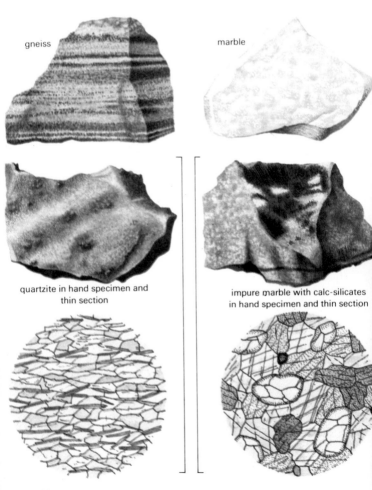

gneiss

marble

quartzite in hand specimen and thin section

impure marble with calc-silicates in hand specimen and thin section

Classification

Classification of metamorphic rocks is less rigidly defined than is the case in the other rock groups. A small number of fundamental rock names are recognized on the basis of general mineralogical and textural features, and for greater precision these may be given prefixes indicating important constituents, as in 'garnet-mica schist', for example. The principal rock names in common use are listed in the table below and overleaf.

ORIGINAL ROCKS	METAMORPHIC ROCK NAMES	GENERAL DESCRIPTION
	slate	Very fine grained with flat, platy cleavage. Often dark in colour. A product of low-grade metamorphism.
argillaceous sediments	phyllite	Fine grained, with undulating, lustrous cleavage planes. Commonly greenish grey. Higher grade than slate.
	mica schist	Medium to coarse grained, with rough, often puckered cleavage parallel to a mica foliation. Middle to high grades.
mixed sediments or acidic igneous rocks	gneiss	Medium to coarse grained, foliated with quartzofeldspathic and micaceous, streaky bands. High grade.
	granofels	Even grained, fine to coarse, unfoliated. Often quartzofeldspathic. Medium to high grade.
arenaceous sediments	quartzite	Dominantly of interlocking quartz grains and often lacking foliation. Occasionally micaceous. Frequently white.
calcareous sediments	marble	Composed of interlocking calcite or dolomite grains, sometimes with calc-silicate minerals. White, grey, or buff.

ORIGINAL ROCKS	METAMORPHIC ROCK NAMES	GENERAL DESCRIPTION
basic igneous rocks	actinolite and chlorite schists	Strongly foliated, green rocks with undulating or puckered cleavage. Low to middle grades.
	amphibolite	Medium- to coarse-grained, nearly black rock. Often foliated and banded. Composed of hornblende and plagioclase.
	eclogite	Medium to coarse grained, with grains of red garnet set in a matrix of green pyroxene. High grade.
various	hornfels	A general name for rocks formed by contact metamorphism. Many are fine grained, dark, and lack foliation.
	mylonite	Very fine-grained, thinly banded, flinty rock. Often dark in colour. The results of mechanical crushing and grinding.

Metamorphism of sedimentary rocks

Among the common sediments, the shales, greywackes, and impure varieties of limestone show extensive changes in both mineralogy and texture as a result of metamorphism whereas orthoquartzites, arkoses, and pure limestones are modified in texture but change little in mineralogy.

The reactivity of shales and greywackes is largely a consequence of the instability of clay minerals at elevated temperatures. Under metamorphic conditions, clay reacts with the other major constituents – quartz, iron oxides, and sometimes lime – to produce a wide variety of aluminous minerals. At the lowest grades, loss of water and reduction of iron and organic material cause hardening and changes in colour, and in orogenic belt metamorphism a slaty cleavage appears early, showing that recrystallization has begun although the grain is at first too fine to allow identification under the microscope. At a more advanced stage, recrystallization of the clay minerals is seen to have created new micaceous minerals such as muscovite and biotite, with chlorite at low grade. In regional metamorphism, this corresponds with a transition in rock type from slate to phyllite and then to schist or gneiss, but in rocks metamorphosed by igneous intrusions, the micaceous flakes lie at random and the rock becomes a tough, dark-coloured hornfels. Quartz and subordinate amounts of feldspar usually accompany the micaceous minerals and garnet often forms porphyroblasts at medium grade and above. One or other of the aluminium silicates may make an appearance in some bands, kyanite or andalusite being typical of medium grades while sillimanite is more common at high grade.

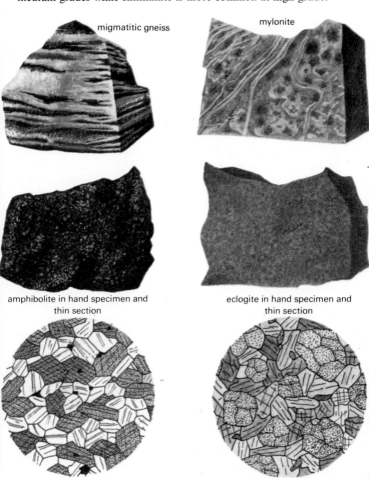

migmatitic gneiss

mylonite

amphibolite in hand specimen and thin section

eclogite in hand specimen and thin section

Metamorphism of pure limestones merely recrystallizes them to white or grey marbles, but if clay minerals, quartz, or other impurities are present, a great variety of new minerals may appear, most of them calcium or calcium magnesium silicates. The pyroxene, diopside, and green to red garnets are particularly common, while in magnesium-rich rocks, needles of the amphibole, tremolite, and small grains of olivine often occur. Most of the metamorphic reactions involve the breakdown of the original carbonates, lime and magnesia combining with the impurities and carbon dioxide being driven off as gas.

Quartz-rich sandstones and arkoses offer few opportunities for metamorphic reactions and change is largely confined to recrystallization into tight-fitting grain mosaics, giving equigranular, very hard, and usually white quartzites.

Metamorphism of igneous rocks

The constituents of acidic igneous rocks are relatively unreactive over a wide range of metamorphic conditions, though at low grade their feldspars may be partly replaced by muscovite mica. Textural changes include the coarsening of originally fine-grained varieties and, in regional metamorphism, the imposition of a moderate degree of preferred orientation, expressed mainly by micas. At higher grades, the textural reorganization can also include segregation of the light- and dark-coloured minerals into the streaky bands characteristic of many gneisses. If very high temperatures are attained, new minerals such as pyroxene may take the place of the biotite mica common at medium grades, and the rocks become darker in colour and less strongly foliated.

Basic igneous rocks display a more varied response to metamorphic conditions. The chief constituents of the original rocks – pyroxene and a calcium-rich plagioclase feldspar – break down at low grade to flakes of chlorite, needles of the amphibole, actinolite, carbonates, and calcium-poor plagioclase. In orogenic belts, these minerals make up dark green phyllites and schists. At somewhat higher temperatures, chlorite and actinolite give way to more robust prisms of dark green hornblende and the rocks appear almost black in hand specimen though often speckled with small, red garnets. This stage corresponds with the formation of garnet-mica schists from argillaceous sediments. At still higher temperatures, pyroxenes take the place of hornblende and the rocks become granular and more massive. If high temperatures are combined with high pressures, however, an unusually dense rock is formed in which the chief constituents are a bright green pyroxene and a deep red garnet, plagioclase being absent. This attractive rock, known as **eclogite** is rather sparsely distributed in orogenic metamorphic belts but is undoubtedly a much more common product of metamorphism in the deep crust and upper mantle.

Polymetamorphism

Many metamorphic episodes affect not only igneous rocks and sediments but also rocks which have already been metamorphosed during some previous event. Rarely will the second metamorphism take place under precisely the same conditions as the first, so that normally there will be textural and mineralogical readjustments to suit the new situation. Rocks metamorphosed more than once are described as **polymetamorphic**, and their recognition depends on the survival of characteristic features imposed during the earlier event. Examples are to be found where the slates and schists of orogenic belts have been invaded by igneous intrusions.

Migmatite outcrop, Norway.

Dynamic metamorphism along faults

Metamorphism resulting from strong shearing stresses set up by deep-seated fault movements produces rocks differing from those so far described in that their textures are at least partly fragmental. In the immediate vicinity of the fault, the zones of severest deformation are picked out by bands of dark-coloured, platy rock called **mylonite**, in which the original mineral grains have been streaked out and crushed to a fine powder and simultaneously fused together. In the less intensely deformed rocks flanking the fault, movement tends to have concentrated along undulating shear planes which slice up the original rock into lens-shaped fragments which internally preserve much of their original texture but are separated by laminae of crushed material.

Ultrametamorphism

The rise in temperature which is one of the chief causes of metamorphism may locally exceed the level at which rocks begin to melt. The boundary between metamorphism and igneous activity which has thus been reached is far from well defined because rocks of different composition begin melting at different temperatures and the melting of any one rock type takes over a range of temperature. Rocks formed under these transitional, or **ultrametamorphic** conditions are exposed to view in the deeply eroded parts of many orogenic

Major structural divisions of the continents, indicating main time divisions of metamorphism.

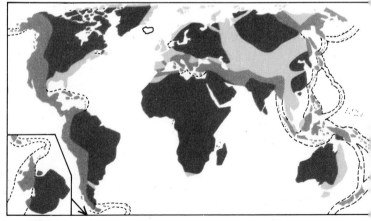

0-200 million years

200-500 million years

more than 500 million years

belts and there form complexes of mixed rocks called **migmatites**. These characteristically contain streaks, bands, and pods of granite more or less enclosed within metamorphic gneisses. The granitic portion of the migmatite represents a low melting point fraction sweated and squeezed out of the heated gneisses. As the process continued, some of the liquid bodies no doubt coalesced and moved upwards to feed granitic intrusions at higher levels in the crust. Orogenic migmatite complexes are one situation in which we can observe the mode of origin of an igneous magma.

Distribution of metamorphic rocks

The principal situations in which metamorphism takes place have been outlined on a previous page, but the present-day geographical distribution of metamorphic rocks deserves further comment. On the continents, such rocks are mainly to be found within eroded orogenic belts because it is here that deep burial and heating were most extensive and here, too, that igneous intrusions are most numerous. Orogenesis has affected every part of the continents at one time or another so that their structure essentially consists of a metamorphosed basement partly covered by a variable thickness of unmetamorphosed sediments. The orogenic belts and their associated metamorphism can be broadly grouped according to their age, as is done in simple fashion in the map opposite, and it is then evident that belts less than about 500 million years old form enormously long, but well-defined zones which in part lie along continental margins, as in the western Americas, but elsewhere traverse the continental interiors, as in the case of the Alpine-Himalayan system. Those parts of the continents not affected by these orogeneses are often referred to as **shields** where the old rocks are exposed at the surface, while similar rocks also extend beneath adjacent areas of undeformed sediments – the rocks of the Canadian Shield beneath the north American prairies, for example. When the shields are examined in detail it is found that they, too, contain deformed and metamorphosed rocks, and are clearly the deeply eroded roots of a plexus of very ancient mountain chains. Metamorphic rocks, then, are present everywhere throughout the continents but are only exposed where a younger sedimentary cover is absent. Those parts of western Europe in which metamorphic rocks can be seen at the surface are shown in the map on page 71.

Although the ocean floors are much less well explored than the continents, it is known that orogenic belts do not occur there, yet it is believed that metamorphism takes place beneath the oceans on at least as extensive a scale as on the continents. Measurements of the rate of heat flowing to the surface and the collection of rocks by dredging indicate that metamorphism mainly takes place beneath the mid-oceanic ridges (map, page 65), although contact metamorphism must also be a common process beneath the numerous central volcanoes which rise from the ocean floor to form islands and sunken **sea mounts**.

ROCK FORMATION AND THE EVOLUTION OF THE EARTH

All mechanisms require energy to drive them and, in the case of the Earth, this comes mainly from the interior, where the breakdown of unstable radioactive elements releases heat. In addition, radiations received from the Sun make a major contribution to the energy requirements of surface processes. In energy terms, no one geological process can be regarded in isolation because each converts the energy it receives into a different form and supplies the power to drive some other mechanism. Regarded in this light the processes which form rocks are part of a natural tendency to re-establish a local equilibrium which has been disturbed by an influx of energy from elsewhere. This is most easily appreciated in the case of metamorphism but is equally true of the igneous activity which occurs when rocks are heated above their melting point and of the erosion

Constructive and destructive plate margins in section.

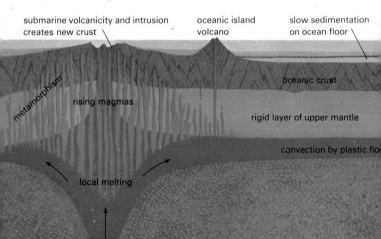

constructive boundary (mid-ocean ridge)

submarine volcanicity and intrusion creates new crust

oceanic island volcano

slow sedimentation on ocean floor

metamorphism

rising magmas

oceanic crust

rigid layer of upper mantle

convection by plastic flo

local melting

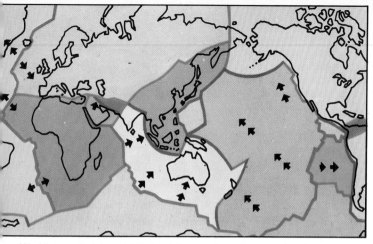

World map of plates and their relative motions. Constructive plate margins shown in red, destructive margins in purple, weakly active boundaries and those along which plates are sliding past each other in green.

and sedimentation which inevitably follow upon the uplift of a mountain chain.

The close correlation between geological events becomes very clear if we consider the geographical distribution of the areas in which rock formation is most active at the present day. For instance, the distribution of modern volcanoes within the continents follows

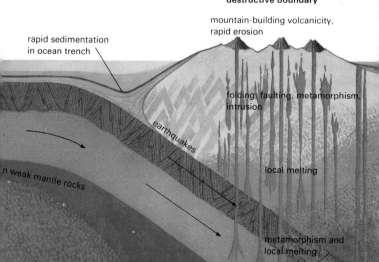

destructive boundary

mountain-building volcanicity, rapid erosion

rapid sedimentation in ocean trench

folding, faulting, metamorphism, intrusion

earthquakes

n weak mantle rocks

local melting

metamorphism and local melting

the pattern defined by the zones of most recent orogenic activity, and the high topographic relief within these zones ensures that erosion and sedimentation are also at their peak in these areas. This high productivity contrasts with the slow rates of erosion and sedimentation, the rarity of earthquakes, and the absence of volcanicity in areas like the Canadian and Baltic Shields. To understand the connections between the rock-forming processes, it is necessary to look more deeply into the structure of the Earth's outer layers.

Recent evidence on the structure of the Earth suggests that the outermost layers of rigid material rest on a much weaker substratum. The rigid zone extends to a depth of between 60 and 100 kilometres and includes not only the rocks of the crust but also part of the upper mantle. The weakness of the rocks below is a consequence of the increase in temperature which at these depths approaches the level of melting and renders the rocks plastic. At even greater depth the very high pressures counteract this effect so that the deeper parts of the mantle are more rigid, though are still able to deform slowly in a plastic manner. If this were an entirely stable state of affairs, the rigid outer layer would remain undisturbed, but in fact, the interior of the Earth is being heated more rapidly by radioactivity than it can be cooled by simple conduction of heat to the surface. Consequently, the mantle is in a more or less constant state of slow convective circulation, the heated rocks being carried up from depth and then spreading out and cooling in the weak zone beneath the outer rigid layer. In accommodating itself to this flowing, plastic substratum, the rigid layer has segmented into a mosaic of separate **plates** which are continuously in motion relative to each other at speeds of up to several centimetres per year. The present-day boundaries of these plates are shown on page 115, some of them lying entirely within the ocean basins and others including both oceanic and continental areas.

The plates are in motion, however, so that this pattern has changed throughout geological time, the segments being pulled apart in some places and pushed hard together in others, creating linear zones of tension and compression along the plate margins. Clearly, such a moving jigsaw puzzle is only possible if the plates are being added to along the zones of tension and gradually destroyed along the zones of compression. Inspection of the map on page 115 will show that addition of new material must take place along the mid-ocean ridges. These great systems of submarine mountains mark the positions of the upwelling currents of hot mantle rock which by spreading laterally push the plates apart and which by partial melting generate the basic magmas which seal the opening fractures as they form. Moreover the heating effect of the upcurrents causes recrystallization in the deeper parts of the newly formed plates, so that the mid-ocean ridges are zones of regional metamorphism as well as of intense igneous activity.

The zones in which plates are destroyed are mostly located along continental margins because here the buoyant effect of the light, continental crust causes plates which contain such material to override denser plates composed entirely of oceanic rocks. Thus, the oceanic plate is forced down into the mantle, its line of descent being marked by a deep sea trench (p 114/5). The enormous frictional drag along the contact of the two plates results in severe folding and metamorphism of the continental margin and of the sediments which overlie the plate boundary, forming an orogenic belt, while

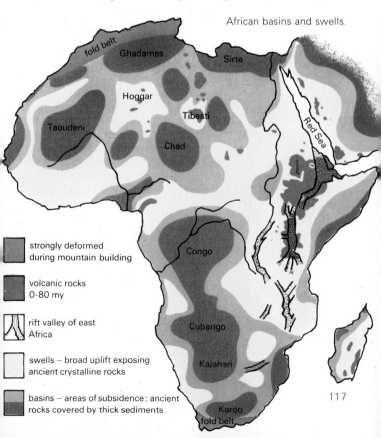

African basins and swells.

strongly deformed during mountain building

volcanic rocks 0-80 my

rift valley of east Africa

swells – broad uplift exposing ancient crystalline rocks

basins – areas of subsidence; ancient rocks covered by thick sediments

117

melting within the descending plate and downbuckled continental rocks is expressed by voluminous igneous activity. Periodically, the folded rocks are uplifted and subject to vigorous erosion, producing an abundance of sedimentary rocks.

Note that whereas the oceanic crust is being steadily created at the mid-ocean ridges and as steadily destroyed beneath the orogenic belts, the lighter continental rocks lose little or any of their substance in the process. It follows that the whole of the oceanic crust is younger than many of the continental rocks, and it has been shown that few parts of the ocean floor contain rocks older than 200 million years whereas the oldest known continental rocks date back to 3500 million years. It is quite possible that the continents have grown continuously larger with time because new material is added to them by igneous activity in the orogenic belts.

While rock formation is most intensive along plate margins, the rate of production within the plates is by no means negligible. Examples of highly productive areas within oceanic plates include the islands of the Hawaiian archipelago, with their abounding volcanicity and aprons of marine sediment, while Africa (p 117) provides good illustrations of activity within a predominantly continental plate. For the last 500 million years or so Africa has been only marginally affected by orogenesis and its structural development within that time has been dominated by broad warping movements which have created extensive basins separated by equally extensive uplifted areas, or **swells**. The swells expose very ancient rocks while the basins are filled with younger sediments eroded from them. Many of the swells are crossed by major fault systems with which volcanicity is often associated. This is particularly apparent on the eastern side of the continent where the fracture systems of a number of swells unite to form a spectacular series of fault-bounded troughs,

Moon rocks as seen under the microscope.

lunar basalt in section

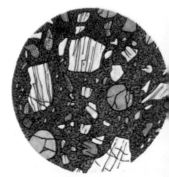

lunar breccia in section

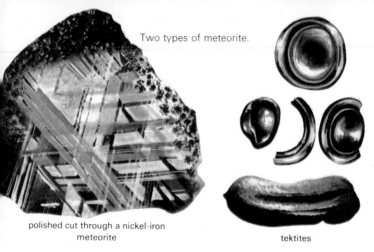

Two types of meteorite.

polished cut through a nickel-iron meteorite

tektites

or **rift valleys**, and a related chain of volcanic fields. The close relationship between swells and igneous activity suggests that mantle activity is responsible for their formation. Possibly the swells form over convectively rising mantle currents similar to those beneath mid-ocean ridges but on a smaller scale.

MOON ROCKS AND METEORITES

Knowledge gained from the American and Russian space programmes makes it possible to compare the rocks of the Earth with those of the Moon. Both these bodies came into existence 4600 million years ago and both soon developed a layered structure, but while the Moon possesses a crust and mantle, its low average density (3.36 g/cm^3) suggests that if a heavy metallic core is also present it must be very small. Moreover, because of its low gravitational attraction the Moon soon lost nearly all its water and atmospheric gases, which on our own planet are so essential to surface processes. Rocks exposed on the Moon were never subject to chemical weathering or to the action of wind, water, or ice, and the most important surface effects have been due to impacting meteorites. Swarms of these small rock fragments are scattered throughout the solar system and are constantly swept up by the planets. The Earth's atmosphere shields it from all but the larger meteorites, smaller ones burning up as 'shooting stars' or disintegrating to dust before they can reach the ground, but there is no such protection on the Moon. Large meteorite impacts account for most of the Moon's craters and the whole surface is covered by a layer of loosely compacted debris made up of fragments thrown out of the craters, together with beads of glass formed by impact melting, and a small amount of meteorite material itself. This process has continued throughout lunar history but radioactive dating of rocks collected on the Apollo missions has shown that around 3900 million years ago impacting reached cataclysmic proportions, and exceptionally large meteorites gouged out huge depressions in the crust, some of them hundreds of kilometres in diameter. Not long after this event, extensive melting in the mantle culminated in eruption of enormous quantities of basic magma similar in composition to terrestrial basalt, but more fluid. Lava flooded the big craters and produced the dark areas we call the lunar maria, or 'seas', a process which continued intermittently until about 3150 million years ago. The paler-toned lunar 'highlands', on the other hand, are composed mainly of primitive crustal rocks and are much more heavily cratered than the maria because of their greater age. Although some granites occur in the crust, the predominant rock type is **anorthosite**, a rock composed largely of plagioclase feldspar.

Specimens of meteorites like those which have had such a formative influence on the landscape of the Moon have often been found on Earth and include varieties composed almost entirely of metallic nickel-iron alloys, while others are made up largely of silicates such as olivine and pyroxene. There are also the curious glass buttons and globules known as **tektites**, which are regarded by some as meteorites though others believe them to be impact-melted ejecta from meteorite craters.

THE ECONOMIC IMPORTANCE OF ROCKS

Even if we exclude the mineral deposits described earlier in this book, the economic uses to which rocks are put are many and various. This is particularly true of sedimentary rocks, some of which are exploited on a very large scale – 50 million tonnes of limestone annually in Great Britain alone, for example.

The economic importance of coal as a source of energy hardly needs emphasizing and it is mined in enormous quantities; world annual production is about 3000 million tonnes of which Britain accounts for over 200 million tonnes, but heavy industry is dependent on rocks in many other ways. Almost all the raw materials of the construction industry, for instance, come from rocks. The manufacture of Portland cement requires sand, clay, limestone, and a little gypsum. More sand and gravel are needed when the cement is used to make concrete, and the steel used in reinforcing the structure may have been smelted from iron-rich sedimentary rocks. Bricks, tiles, and other ceramics are made from mixtures of various types of clay or shale, mortar and plaster are prepared from hydrated lime and gypsum, and the making of window glass requires sand of high purity.

The direct use of rocks as building stones is less common than it used to be because of its great expense compared with concrete or brick, but good stone is still in demand for high quality construction and as cladding for buildings made of concrete. A first-class building stone must reach a high specification and it is not surprising that such material can command a price high enough to make it worth transporting long distances or even exporting. Many of the best-known building stones are granites, but syenites and some orthoquartzite sandstones are also good, while rocks such as mottled slates find an increasing use as decorative slabs. In Europe, limestones have been employed for building perhaps more often than any other rock because their appearance is attractive and their cost relatively low, but unfortunately they are only moderately resistant to weathering, especially in industrial areas. All these rock types and many others are crushed in large quantities to make roadstone, ballast, and similar aggregates.

Opencast working of sedimentary ironstone, Northamptonshire, England.

It is easy to find other instances of our dependence on rocks. The chemical industry uses large amounts of limestone and evaporite sediments as basic raw materials while, apart from exploiting sedimentary ores, the British steel industry annually consumes the 7 million tonnes of limestone added to the blast furnaces to form slag with the impurities. A comparable amount of limestone is used in agriculture for improving heavy soils, and this industry is also the principal consumer of potassium-bearing evaporites and phosphate rocks essential for maintaining soil fertility.

As the world's reserves of rich mineral deposits run out, increasing amounts of the essential metals will have to be obtained from the lower concentrations found in common rock types. Already copper can be economically worked in deposits containing less than 1 per cent of the metal, provided that there are large volumes of ore available. Some common rocks contain large amounts of useful metals but at present cannot be exploited because techniques to extract the metals profitably have not yet been found. A notable example is aluminium, the third most abundant element in the Earth's crust and present in amounts between 15 and 20 per cent in ordinary shales and clays.

MAKING A COLLECTION OF ROCKS

If you wish to collect rocks for yourself, you would be well advised to find out first what geological maps and guides are available covering the area in which you are interested. In most countries the government department responsible for geological surveys publishes a more or less complete cover of maps, the main series being normally on a scale 1:50 000 or 1:100 000. Areas of special interest may also be covered on larger scales, while usually there are small-scale maps covering the whole country or a large part of it on one sheet. In Britain, the Institute of Geological Sciences publishes a wide range of maps (though not all are in print at the present time), and these can be obtained from the Geological Museum in London, the various branches of Her Majesty's Stationery Office, or through a book seller. The Institute also publishes an excellent series of handbooks of regional geology which covers England, Wales, and Scotland in eighteen parts. For descriptions of individual exposures you may need to consult the more detailed Memoirs, but good excursion guides are published by a number of local geological societies and by the Geologists' Association.

Little equipment is needed: a geological hammer at least 1 kilo in

In the Geological Museum, South Kensington, London.

weight, a small cold chisel, a black felt-tip pen for numbering rocks in the field, newspaper for wrapping specimens, and a field notebook. Permission should always be sought to collect on private land and collecting anywhere should be done with restraint. Indiscriminate use of hammers and chisels is a potent agent of erosion, and landowners are rightly incensed by the careless undermining of walls, river banks, and the like. There is often sufficient loose rubble at an exposure to make the hammering of undisturbed rock unnecessary, but it is important to obtain unweathered material and fallen blocks may have to be split open. The most useful size of hand specimen measures about 10 by 8 by 3 centimetres. The field notebook is used to record the precise locality, giving a map reference if possible, together with the number of the specimen and a full description of all the rocks seen in the exposure.

Later, the specimens may be permanently labelled by painting small rectangles of white enamel on them, writing on the paint with Indian ink and protecting the writing with a coat of varnish. Specimens are best stored in cardboard trays within shallow drawers.

Help in identifying specimens may often be obtained from members of the local geological society, departments of geology in universities and polytechnics, or from the Geological Museum in London.

BOOKS TO READ

Identification guides (containing illustrations in colour)

Boegel, H. (Edited and revised by H Sinkankas). 1971. *A Collector's Guide to Minerals and Gemstones*. London, Thames and Hudson.

Hamilton, W R, Woolley, A R, and Bishop, A C. 1974. *The Hamlyn Guide to Minerals, Rocks and Fossils*. London, Hamlyn.

Kirkaldy, J F. 1970. *Minerals and Rocks in Colour*. London, Blandford.

General references for minerals and rocks

Battey, M H. 1971. *Mineralogy for Students*. London, Longman.

Cox, K G *et al*. 1974. *Practical Study of Crystals, Minerals and Rocks*. Maidenhead, McGraw-Hill.

Hatch, F H *et al* 1971. *Petrology of the Sedimentary Rocks*. London, Allen and Unwin.

Hatch, F H *et al.* 1973. *Petrology of the Igneous Rocks*. London, Allen and Unwin.

Read, H H. 1970. *Rutley's Elements of Mineralogy*. London, Allen and Unwin.

(Unfortunately, there is no straightforward text-book dealing with metamorphic rocks readily available: see *Introduction to Geology* below for a useful short account. Harker's *Metamorphism* is now rather out-of-date and of limited value.)

General geology references

Holmes, Arthur. 1965. *Principles of Physical Geology*. London, Nelson.

Read, H H and Watson, J. 1966. *Beginning Geology*. London, Macmillan.

Read, H H and Watson, J. 1970. *Introduction to Geology* (Volumes 1 and 2), London, Macmillan.

Smith, A J. 1974. *Geology*. London, Hamlyn.

Whitten, D G A and Brooks, J R V. 1972. *The Penguin Dictionary of Geology*. Harmondsworth, Penguin Reference Series.

Guides

Institute of Geological Sciences. *British Regional Geology*. London, Her Majesty's Stationery Office. A series of nineteen guides covering Britain and Northern Ireland. The most useful source of general geological information about individual British areas. Many other guides are published by local societies and the Geologists' Association. The address of the Secretary of the Geologists' Association can be obtained from the current part of the *Proceedings* which any library can obtain.

Collecting equipment

Equipment for collecting can be obtained from Gregory Bottley and Company, 30 Old Church Street, London SW3. They also sell specimens and welcome visitors at their showrooms at the above address.

INDEX

Page numbers in **bold** type refer to illustrations